Bioinspired Nanomaterials for Energy and Environmental Applications

Edited by

Alagarsamy Pandikumar[1] and Perumal Rameshkumar[2]

[1]Electrochemical Materials Science, Functional Materials Division, CSIR-Central Electrochemical Research Institute, Karaikudi-630003, Tamil Nadu, India

[2]Department of Chemistry, Kalasalingam Academy of Research and Education, Krishnankoil, 626 126, Tamil Nadu, India

Copyright © 2022 by the authors

Published by **Materials Research Forum LLC**
Millersville, PA 17551, USA

Published as part of the book series
Materials Research Foundations
Volume 121 (2022)
ISSN 2471-8890 (Print)
ISSN 2471-8904 (Online)

Print ISBN 978-1-64490-182-3
eBook ISBN 978-1-64490-183-0

Distributed worldwide by

Materials Research Forum LLC
105 Springdale Lane
Millersville, PA 17551
USA
https://www.mrforum.com

Manufactured in the United States of America
10 9 8 7 6 5 4 3 2 1

Table of Contents

Preface

Recent Advancement in Green Synthesis of Metal Nanoparticles and their Catalytic Applications
P. Arunkumar, S. Saran, G. Manjari and K. Mohanty 1

Bio-Inspired Metal Oxide Nanostructures for Photocatalytic Disinfection
Muthuraj Arunpandian, Tammineni Venkata Surendra, Norazriena Yusoff,
Saravana Vadivu Arunachalam 39

Bioinspired Nanomaterials for Photocatalytic Degradation of Toxic Chemicals
Vellaichamy Balakumar, Ramalingam Manivannan and Keiko Sasaki 83

Bioinspired Nanostructured Materials for Energy-Related Electrocatalysis
M. Rajkumar, C. Pandiyarajan and P. Rameshkumar 117

Bioinspired Nanomaterials for Supercapacitor Applications
Adhigan Murali, R. Suresh Babu, M. Sakar, Sahariya Priya, R. Vinodh,
K. P. Bhuvana, Senthil A. Gurusamy Thangavelu, M. Abdul Kader 141

Bio-Mediated Synthesis of Nanomaterials for Dye-Sensitized Solar Cells
G. Murugadoss, T.S. Shyju, P. Kuppusami 175

Bioinspired Synthesis of Nanomaterials for Photoelectrochemical Applications
M.L. Aruna Kumari 211

Keyword Index

About the Editors

Preface

Nanomaterials with its numerous advantages made the research community to explore them with different biomolecules for making them further applicable in different fields. Bioinspired synthetic route provides a nontoxic and reliable pathway for preparing nanomaterials with different size, shape, composition and physicochemical properties. Bio-scaffolds as templates in forming nano-bio heterojunctions will be advantageous in applications from catalysis, sensors to energy related applications like electrocatalysis. The biomediated nanomaterials have significantly contributed towards a variety of applications. Thus, detailed information on the bioinspired nanomaterials and their key applications must be obtained in order to realize their high potential to the maximum. The introduction of bioinspired nanomaterials along with some miscellaneous applications has already been discussed in an earlier volume "Bioinspired Nanomaterials - Synthesis and Emerging Applications", hence, the present book aims to discuss elaborately on catalytic applications of bioinspired materials including photocatalysis, electrocatalysis and photoelectrocatalysis. Further, individual chapters are allocated for supercapacitor and solar cells applications. The recent progress in developing bioinspired composites and the perspectives on future opportunities in various catalytic applications is also discussed in this book.

This book covers the importance of bioinspired synthesis of nanomaterials and their energy and environment related catalytic applications. It also covers the applications of bioinspired materials in supercapacitors and solar cells. The audience can easily understand the role of bioinspired nanomaterials in the current scenario of energy and environment relevant applications oriented research and development. The book will be highly helpful for researchers to establish their own research in the area of bioinspired nanomaterials.

We are very much grateful to all the authors who contributed their chapters to make a valuable book and for the successful completion of the process. We are thankful to the editor, Thomas Wohlbier, Materials Research Forum LLC for accepting our proposal and giving an opportunity to edit this book and his help towards the successful completion of the work is greatly acknowledged.

Dr. Alagarsamy Pandikumar (Leading Editor)

Scientist, Electro Organic and Materials Electrochemisry Division, CSIR-Central Electrochemical Research Institute, Karaikudi-630003, Tamil Nadu, India

Dr. Perumal Rameshkumar (Associate Editor)

Assistant Professor, Department of Chemistry, School of Advanced Sciences, Kalasalingam Academy of Research and Education, Krishnankoil, 626 126, Tamil Nadu, India

Bioinspired Nanomaterials for Energy and Environmental Applications   Materials Research Forum LLC
Materials Research Foundations **121** (2022) 1-38   https://doi.org/10.21741/9781644901830-1

Chapter 1

Recent Advancement in Green Synthesis of Metal Nanoparticles and their Catalytic Applications

P. Arunkumar[1#], S. Saran[2#], G. Manjari[3] and K. Mohanty[2,4]*

[1]Department of Ecology and Environmental Sciences, Pondicherry University, Puducherry-605014, India

[2]Centre for the Environment, Indian Institute of Technology Guwahati, Guwahati - 781039, India

[3]Department of Ecology and Environmental Sciences, Pondicherry University, Puducherry-605014, India

[4]Department of Chemical Engineering, Indian Institute of Technology Guwahati, Guwahati - 781039, India

*kmohanty@iitg.ac.in, #Equal contribution

Abstract

Nanotechnology is growing as an essential discipline in research with several applications. For the synthesis of nanoparticles, distinct methods such as chemical, physical and biological entities have been adopted. Among these, green synthesis of metal nanoparticles has gained wider attention owing to its simplicity, non-toxicity, cost-effective, sustainable and easy availability. The phyto compounds existing in the plant extract either acts as reducing, stabilizing and capping agents; thereby, it reduces the cost and consumptions of energy and hazardous chemicals. This review mainly focused on the recent trend in the utilization of the plant material for the synthesizing of various metal nanoparticles and their catalytic application with special reference to their morphological features.

Keywords

Biogenic Synthesis, Metal Nanoparticles, Phytochemicals, Catalysis, Organic Pollutant Degradation

Contents

Recent Advancement in Green Synthesis of Metal Nanoparticles and their Catalytic Applications...1

1.   **Introduction**...2

**2.   Biogenic synthesis of silver nanoparticles and their
catalytic applications**...6

**3.   Bioinspired synthesis of gold nanoparticles and their catalytic
applications**...11

4.   Bioinspired copper nanoparticles in catalysis..15

**5.   Biomediated synthesis of palladium nanoparticles and their
catalytic applications**...15

**6.   Bio-fabrication of platinum nanoparticles and their catalytic
applications**...17

**7.   Green synthesis of iron nanoparticles and their
catalytic applications**...19

**8.   Mechanism of catalytic conversion of 4- Nitrophenol to
4- Aminophenol using green synthesized metal nanoparticles**....................23

Conclusion...24

Acknowledgement..24

Conflict of interest...24

References..25

1.   Introduction

Nanotechnology has deep-rooted itself as the emerging front-line technology of scientific alma matter for controlling the size at a molecular level. Richard Feynman, the pioneer of nanotechnology, introduced the theory of nanotechnology in his famous discourse quoting "There's plenty of room at the bottom" in the year 1959. The word Nano popularized by Norio Taniguchi is derived from Greek etymology defining the word as the dwarf, too small or tiny, referring to the size in one billion [1]. The intriguing and fascinating nature has rendered remarkable and game-changing applications in optics, electronics, catalysis, biotechnology, environment etc. Nanoparticle synthesis with controlled size and shape produces a conspicuous response on their physical properties with reduced imperfections, spatial confinement, greater surface energy and larger surface area [2].   Basically, nanoparticle synthesis involves top down and bottom up approach using biological, chemical and physical means. In the top down approach, the bulkier materials are cracked down into nanosized materials by adopting chemical and physical synthesis like electrodeposition [3], laser [4], vaporization and vacuum deposition [5], mechanical attrition [6], precipitation, sol-gel techniques. Whereas, in the bottom up approach, molecules or atoms assemble to pile up structures in nanometer range by manifesting chemical methods adopting chemical methods utilizing chemicals and solvents [7, 8] and biological synthesis was utilizing biological entities like plants, bacteria, fungus, yeast, algae etc. [9]. With mounting environmental concerns, green chemistry principles were applied for the sustainable nanoparticle synthesis without the involvement of any toxic solvents, high energy and chemical consumption. Though synthesis of metal nanoparticle synthesis via physiochemical means is an intriguing area of research, the scientific fraternity is anxious towards the outcome like usage of toxic solvents, hazardous by-products and higher energy intake [10]. To weed out these disadvantages, biomimetic synthesis of materials at the nanoscale level using biological organisms were preferred. The need for a benign, alternative and eco-friendly synthesis of nanoparticles can be accomplished by utilizing different natural resources. Various phytochemicals present in the biological entities such as flavonoids, polyphenols, terpenoids, proteins participate in the metallic nanoparticles synthesis by reducing the precursor salts and stabilize the nanoparticles in a complicated process. The challenging part of the green synthesis lies in exactly identifying the chemical constituent involved the formation of metal nanoparticles. Schematic representation of the metal nanoparticles synthesis using different methods, as illustrated in Figure 1 [12].

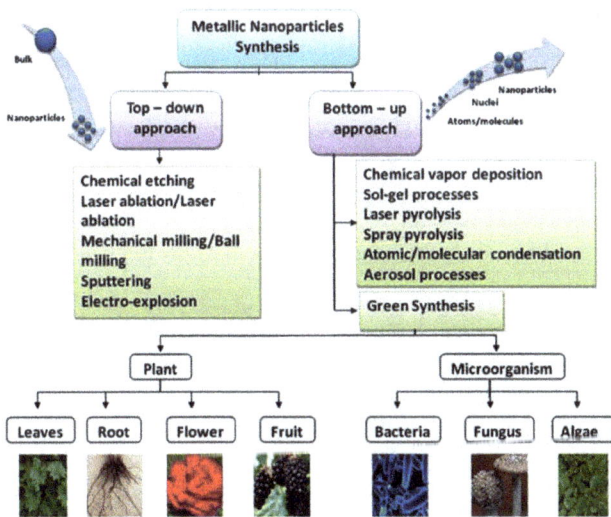

Figure 1. Schematic representation of the synthesis of metal nanoparticles using different methods. Reproduced permission from Ref. No. [11]

The production of metal nanoparticles involves three major attributes viz. solvent medium, stabilizing agent and reducing agent. In green synthesis, the phytochemicals serve as both reducing and stabilizing agent. The phytochemicals being such diverse makes it an annoying task to pin down a particular chemical constituent accountable for reducing and stabilizing metal nanoparticles. Phytochemicals comprising polyphenols, flavonoids, terpenoids, phenolic acid, proteins, organic acid are considered to be responsible for reducing and stabilizing agents in the nanoparticle synthesis [12, 13]. Various phenolic acids such as gallic acid, protocatechuic acid, caffeic acid and ellagic acid are utilized in metallic nanoparticle synthesis as reducing agent. Phenolic acid possesses the metal chelating ability and antioxidant potential, and it belongs to the polyphenol family[14-16]. It is surmised that silver nanoparticles (Ag NPs) are produced through the development of intermediate complex of silver ions with phenolic hydroxyl moieties present in the gallic acid. Different plant genera form mesophyte to xerophyte are known to possess metal-reducing potential. Metal ion reduction is enhanced by the release of active hydrogen from hydrophytes owing to the presence of catechol, ascorbic acid and protocatechuic acid in them [17]. To ascertain this, the involvement of purified phenolic like protocatechuic acid and gallic acid in capping and bio-reducing of metals [18, 19]. Flavonoids are the important constituents of plant phytochemicals found in different plant parts like fruits, flower skin, the epidermis of the leaves for protection against adverse environmental conditions. Due to their diverse structural and chemical properties, these secondary metabolites are found in diverse forms like flavonols, anthocyanidins, flavonones, flavones and isoflavones. Flavonoids with hydroxyl group are hydrophilic, whereas the flavonoids with isopentyl and methyl groups are lipophilic. Zhou et al., 2010 [20] reported that flavonoids are the major phytoconstituents linked with reducing potential by donating electron or hydrogen atoms. Ahmad et al., 2010 [21] proposed that flavonoids like rosmarinic acid and luteolin produce free hydrogen in the course of flavonoid to ketoenol conversion by reducing silver ions into silver nanoparticles. Similarly, Ghoreishi et al., 2011 [22] reported that during the reduction of metal ions, the hydroxyl group of the flavonoids are oxidized into carbonyl groups.

Both the crude and purified form of flavonoids are reported to synthesize metal nanoparticles. Sahu et al., 2016 [23] used to prepare Ag NPs from the purified form of naringin, diosmin and hesperidin. The silver ions were reduced to nanoscale particles by utilizing poly hydroxylated secondary metabolites of the plant by acting as the capping agents. Also, gold nanoparticles were formed either by electrostatic force or ionic bond with a significant linear relationship between antioxidant activity and reducing potential [20]. Ovasis et al. 2018[16] reported that the release of atomic hydrogen during the tautomeric conversion of flavonoid is responsible for reducing metallic ions into nanoparticles. Trouillas et al., 2006 [24] revealed in DFT studies that flavonoids possessing hydroxyl group with bond breaking power involve themselves as reducing and capping agent during photosynthesis. Besides, Yoosaf et al., 2007 [25] observed that polyphenols contain no less than two hydroxyl functions at the second or fourth position exhibit excellent reducing ability.

Terpenoids are the phytochemicals responsible for taste, aroma and color in plants. They are the secondary metabolites produced in the plant-derived from the essential oils containing terpenoids, sesquiterpenes, terpenes, and oxygenated compounds like ketones, lactones, phenols, aldehydes, ether alcohols, esters, and phenolic ethers [26]. Kim 2009[27] reported that sesquiterpenoids and monoterpenoids play a crucial role in silver nanoparticle synthesis as donors. Besides, various plant species have reported for silver nanoparticle synthesis like *Anacardiumoccidentale*[28], *Ricinuscommunis* [29], *Myristicafragrans* [30], *Cocos nucifera* [31]. Similarly, the hydroxyl groups, monoterpenes, terpenoids like linalool, methyl chavicoland eugenol also exhibit potential metal reduction [32]. In phytomediated nanoparticle synthesis, the proteins also play the role of reducing agent by donating electrons and subsequent stabilization for the nanoparticle formation. Li et al., 2007 [33] revealed the action of proteins in Ag NPs synthesis by utilizing the extract of *Capsicum annum.*

The UV absorption studies revealed the peptide bond presence around 280 nm, indicating the presence of amino acids and their interaction with silver ions. Transmission electron microscopy investigations revealed the particles were attached through proteins. Also, cyclic voltammetry indicated the reductive aminepresence, which becomes oxidized during silver ion reduction by gaining electrons and subsequent nucleation of silver to form particles in nanoscale. The role of proteins during phytomediated nanoparticle synthesis was proved by Mukherjee et al., 2014 [34]. Their results proved that aqueous extract contained low molecular weight proteins but was absent in the synthesized nanoparticles.

Similarly, Shukla et al., 2008 [35] reported the phytomediated gold nanoparticles (Au NPs) from the soya bean extract. Recent researche have manifested the role of sugars in the formation of nanoparticles as stabilizing and reducing agents [36, 37]. The role of glucose, fructose, hemicellulose was explained in the biosynthesis of Ag NPs in which glucose was reported for higher reducing capacity whereas, the fructose was attributed with limited reduction potential owing to their limited tautomeric shift and antioxidant potential [38]. In addition to plant-mediated biosynthesis of nanoparticles, microbial communities like bacteria and actinomycetes are also reported as the proficient biological entities by reducing metal ions into metallic nanoparticles. The biosynthesis of nanoparticles is categorized into two categories, namely intracellular and extracellular. The extracellular synthesis of nanoparticles especially metal nanoparticles is proven to be economically

Bioinspired Nanomaterials for Energy and Environmental Applications   Materials Research Forum LLC
Materials Research Foundations **121** (2022) 1-38   https://doi.org/10.21741/9781644901830-1

feasible, easy genetic manipulability and elevated growth rate. Bacterial species like *Bacillus amyloliquefaciens, Pseudomonas stutzeri, Acinetobactercalcoaceticus, Bacillus licheniformis, Escherichia coli, Bacillus megaterium* and *Lactobacillus* sp. are reported with the biosynthesis of metal nanoparticles [39]. Schematic representation of the Biogenic synthesis of metal nanoparticles using various plant parts was displayed in Figure 2.

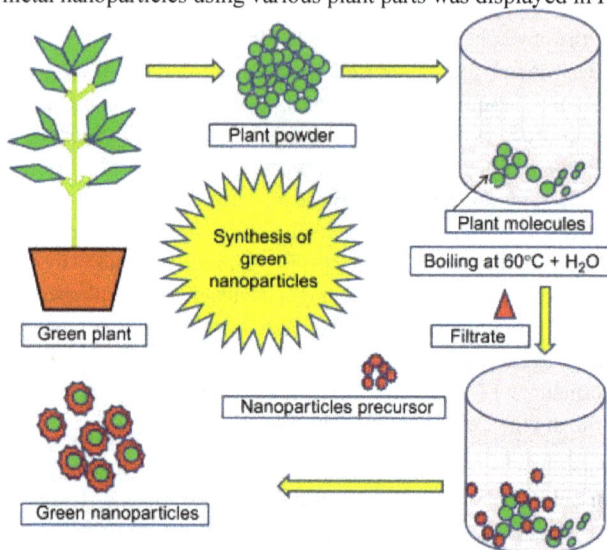

Figure 2. Schematic representation of the green synthesis of metal nanoparticles using plant extract. Reproduced permission from Ref. No. [12]

2. Biogenic synthesis of silver nanoparticles (Ag NPs) and their catalytic applications

Chemical reduction is the most followed method for silver nanoparticles synthesis. Numerous inorganic and organic reducing agents are utilized in aqueous and non- aqueous medium for its synthesis viz. polyethylene glycol, Tollen's reagent, sodium citrate, sodium borohydride, Ascorbate, N, N-dimethyl formamide, essential hydrogen [40-42]. Though, the chemical synthesis produced a considerable amount of nanoparticles, the chemicals utilized are non-ecofriendly and toxic to the environment. Hence, there is a need for synthesizing nanoparticles via green route without using toxic chemicals and developing green protocols using biological entities such as plants, microorganisms, biomass for the existing physicochemical methods [43]. With the advantage of also being cost-effective, the disadvantages like higher temperature, energy, pressure and hazardous chemicals can be weeded out by green synthesis [44].

Francis et al., 2017 [45] exploited the catalytic property of the Ag NPs produced via microwave aided biogenic synthesis by utilizing the extract of *Mussaendaglabrate* leaves. As produced Ag NPs degraded the toxic dyes such as methyl orange and Rhodamine B and in the reduction of 4- Nitrophenol. Joesph and Mathew, 2015 [46] reported microwave aided green synthesis of Ag NPs *Biophytumsensitivum* as stabilizing and reducing agent. The average diameter of the spherical Ag NPs reported in the study was 19.06 nm, and they successfully demonstrated the reduction of methylene blue and methyl orange using sodium borohydride. Jothi and Singh, 2016 [47] described the Ag NPs synthesis using *Zanthoxylumarmatum* leaves with size ranging between 15 - 50 nm. They were proved to be excellent in degrading hazardous dyes like methyl orange, methylene blue, methyl red and saffranin O. Saha et al., 2017 [48] demonstrated biogenic Ag NPs synthesized from the medicinal plant called *Gmelinaarborea* and their application in the reduction of methylene blue dye. The prepared Ag NPs were crystalline in nature of size ranging between 8 to 32 nm, and the catalytic degradation was completed within ten minutes of the experiment. *Amaranthusgangeticus* Linn leaf extracts were utilized for producing the biogenic silver nanoparticles in an aqueous medium. The eco-friendly process resulted in Ag NPs with particle size ranging between 11-15 nm. The biogenic silver nanoparticles exhibited efficient catalytic degradation of Congo red in presence of $NaBH_4$ [49]. Vidhu and Philip, 2014 [50] demonstrated the biogenic silver nanoparticles preparation by utilizing *Trigonellafoenum-graecum* seeds and tested their potency in degrading synthetic dyes by sodium borohydride reduction. Mata et al., 2015 [51] validated the biogenic synthesis of silver nanocatalysts using *Plumeria alba* flower extract by the presence of functional groups like C-C and C-N in the extract. They exhibited efficient catalytic activity by the reduction 4- Nitrophenol into 4-Aminophenol under eight minutes. The as-prepared silver nanoparticles also exhibited moderate degrading efficiency towards ethidium bromide and methylene blue. Veisi et al., 2018 [52] demonstrated bioinspired silver nanoparticle synthesis using *Thymbraspicata* leaf extract without any stabilizer or surfactants. The as-synthesized nanocatalysts exhibited a powerful reduction of toxic dyes like Rhodamine B, 4- Nitrophenol, methylene blue at room temperature in $NaBH_4$ presence by following pseudo-first-order reaction. Varadavenkatesan et al., 2016 [53] established the catalytic efficiency of the silver nanocatalysts produced via green synthesis of *Mussaendaerythrophylla*. These Ag NPsare highly stable with the zeta potential value of -47.7 mV which highlighted the catalytic degradation of methyl orange with sodium borohydride as reductant. Ping et al., 2018 [54] focused on the biogenic silver nanoparticle synthesis and their utility in catalytic degradation of a hazardous dye, Direct Orange 26.

Bioinspired Nanomaterials for Energy and Environmental Applications   Materials Research Forum LLC
Materials Research Foundations **121** (2022) 1-38   https://doi.org/10.21741/9781644901830-1

They also observed that the reaction temperature of the green synthesis played a decisive role in the reduction of Direct Orange 26 by $NaBH_4$ reduction. Optimization of temperature played a central role in the biogenic synthesis of Ag NPs as the temperature above 40 °C had a contra dictory effect on the nanoparticle size with an increase in temperature augmented the particles bigger and hence curtailed the surface catalysis effect. Junejo et al., 2014 [55] synthesized Ag NPs derived from ampicillin in an aqueous medium with crystalline nature exhibiting powerful catalytic activity towards the methylene green reduction. The as-prepared Ag NPs accomplished complete reduction within 4 minutes exhibiting enhanced catalytic potential with easy recovery and reuse. Qing et al., 2017 [56] successfully synthesized Ag NPs are utilizing the water extract of waste tea as stabilizing and reducing agent. The as-prepared biogenic Ag NPs exhibited efficient, rapid and durable catalytic activity towards the cationic organic dye degradation but not with the organic dye of anionic nature at room temperature. The catalytic degradation of cationic organic dye degradation followed a pseudo-second-order model. Francis et al., 2018 [57] fabricated biogenic Ag NPs using *Elephantopusscaber* extract in a domestic microwave oven. TEM studies revealed that the biogenic Ag NPs are spherical shaped with particle size of 37.86 nm. The prepared Ag NPs reduced organic nitro compounds like 4-nitroaniline, 4-nitrophenol and2-nitroaniline and also, eosin Y degraded within a short time. Arya et al., 2017 [58] employed the barks of *Prosopisjuliflora* to produce spherical shaped Ag NPs of size ranging from 10-50 nm. The biogenic Ag NPs also exhibited dye degradation of ninety percent under eighty-minute activity against the carcinogenic dye, 4- Nitrophenol. Edison et al., 2016 [59] fabricated biogenic Ag NPs from the extract of *Terminalia cuneata* barks with a distorted spherical shape of size ranging between 25–50 nm. The biogenic silver nanoparticles showed a promising reduction of direct yellow-12 under sodium borohydride. Ahmed et al., 2014 [60] synthesized biogenic Ag NPs from the compound β-sitosterol-D-glucopyranoside obtained from *Desmostachyabipinnate* with natural sunlight as the initiator through free radical exchange phenomenon by the involvement of reactive oxygen species. These biogenic AgNPs exhibited promising catalytic activity against hazardous carcinogenic dyes like Congo red, acridine orange, methyl red and methylene blue in an aqueous medium. Bogireddy et al., 2016 [61] fabricated crystalline, spherical shaped Ag NPs from Sterculia *acuminata* fruits extract with particle size of ~10 nm. The biogenic Ag NPs were evaluated for the degradation of hazardous dyes like phenol red, methylene blue, methylene orange and 4- Nitrophenol. The degradation efficiency of methylene blue and direct blue 24 was faster than the other dyes. Edison et al., 2016 [62] manifested the reduction/ degradation of carcinogenic azo dyes like Congo red and methyl orange with biogenic silver nanoparticles derived from *Anacardiumoccidentale* extract. Saravanan et al., 2017 [63] reported the silver nanoparticle synthesis from the exopolysaccharide produced from the strain of Leuconostoc *lactis* isolated from the idli batter. EPS stabilized Ag NPs established efficient degradation of dyes like Congo red and methyl orange. Ahmed et al., 2015 [64] demonstrated green Ag NPs synthesis by exploiting redox protein comprised of ferredoxin, ferredoxin NADP$^+$ reductase in natural sunlight. The prepared Ag NPs exhibited promising catalytic activity against hazardous dyes like methyl orange, methyl red and methylene blue. The XRD and TEM image of green synthesized silver and gold nanoparticles were illustrated in Figure 3.

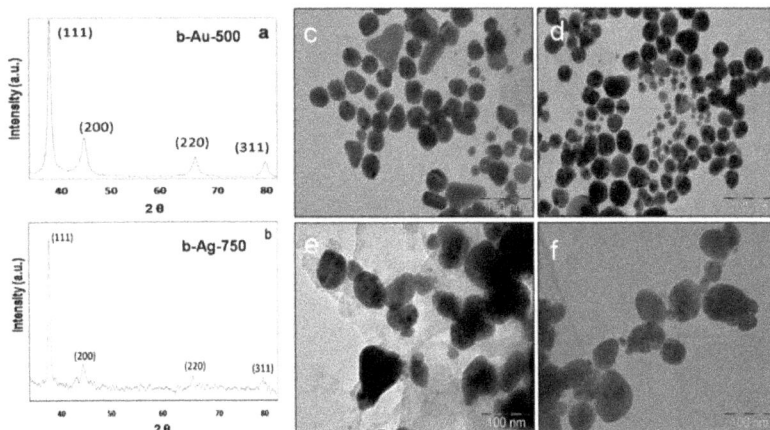

Figure 3. XRD of (a) Ag NPs and (b) Au NPs; TEM image of green synthesized (c&d) Silver and (e&f) Gold nanoparticles. Reproduced permission from reference no [13]

Anand et al., 2017 [65] reported in vitro synthesis of biogenic Ag NPs by exploiting *Ekebergiacapensis* water extract. The prepared Ag NPs with the size ranging from 20-120 nm and were found to have promising catalytic activity against Allura red and congo red by means of electron relay effect attributed with time. Muthu and Priya, 2017 [66] manifested the effective catalytic degradation of toxic dye like 4- Nitrophenol and methyl orange by biogenic Ag NPs prepared from *Cassia auriculata* flowers. The aqueous extracts of the flower acted as reducing and stabilizing agent producing spherical and triangular-shaped silver nanocatalysts with size ranging from 10 to 35 nm. A facile *Aglaiaelaeagnoidea* flower extract facilitated green synthesis of Ag NPs was reported by Manjari et al., 2017 [67]. The prepared particles were about 17 nm in size and exhibited ultra-rapid catalytic degradation of various dyes such as methylene blue in 5 seconds, congo red in 10 seconds and nitrophenol in 15 min (Figure 4, 5 and 6). Vijayan et al., 2019 [68] reported a simple and one-pot green Ag NPs synthesis using the *Nervaliazeylanica* leaf extract by microwave-assisted strategy. The as-prepared Ag NPs with particle size

34.2 nm exhibited potential catalytic activity against organic dyes following pseudo-first-order kinetics.

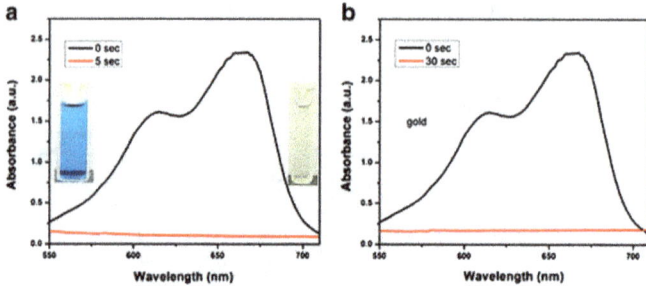

Figure 4. Catalytic reduction of Methylene blue dye using (a) Ag and (b) Au NPs prepared using plant extract. Reproduced permission from Ref. No. [67]

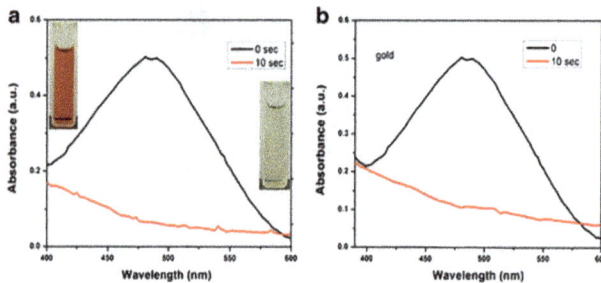

Figure 5. Catalytic reduction of Congo red dye using (a) Ag and (b) Au NPs prepared using plant extract. Reproduced permission from Ref. No. [67]

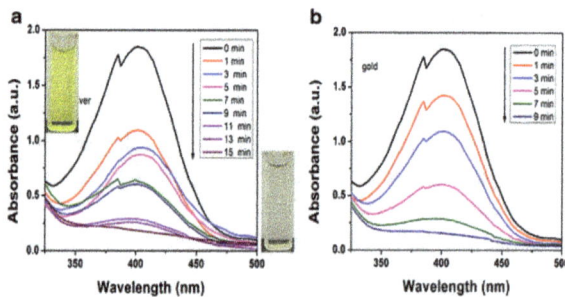

Figure 6. Catalytic reduction of Nitrophenol dye using (a) Ag and (b) Au NPs prepared using plant extract. Reproduced permission from Ref. No. [67]

3.   Bioinspired synthesis of gold nanoparticles and their catalytic applications

Aromal and Philip, 2012 [69] synthesized gold nanoparticles from fenugreek (*Trigonellafoenumgraecum*) performed as both reducing and protecting agent. The active phytochemicals present in fenugreek extracts such as alkaloids, proteins, saponin played a decisive role in the formation of Au NPs. The prepared Au NPs were ranging from 15-25 nm in size. The prepared Au nanoparticles reduced 4 Nitrophenol (4NP) to 4 Aminophenol (4AP) in the presence of excess $NaBH_4$. The result revealed that within 7 min complete conversion of 4NP to 4 AP was observed. The catalytic potency of the gold nanoparticles depended on their size, smaller the size showing the quicker conversion of 4NP to 4 AP.

Gold nanoparticles synthesis was synthesized from fruit extract of *Prunusdomestica* and their catalytic reduction of 4 NP was demonstrated by Dauthal and Mukhopadhyay, 2012 [70]. FTIR studies showed that amino acids and polyols of plum fruit extract function as reducing and stabilizing agent for Au NPs formation. The prepared Au NPs was about 20 nm in size and the reduction of 4NP to 4 AP was finished within 20 minutes by following pseudo-first-order kinetics reaction. Likewise, Maity et al., 2012 [71] demonstrated the biogenic gold nanoparticles synthesis from hetero-polysaccharide, extracted from katira gum. The average size of the synthesized Au NPswas 7 nm and showed a potential catalytic activity towards the reduction of 4 NP to 4 AP in 30 min.

Sen et al., 2013 [72] reported the biogenic synthesis of Au NPs extracted from glucan of an edible mushroom *Pleurotusflorida*. The physical properties like size and shape of the Au NPs are dependent on the concentration of gold precursor solution used in the synthesized process. The particle size increased in size from 5 nm to 15 nm, with an increase in the concentration of the initial gold solution. The glucan compound acted as both stabilizing and reducing agent in the synthesis of Au nanoparticle. The resulting Au NPs reduced 4-NP to 4-AP within 30 min in $NaBH_4$ presence. Majumdar et al., 2013 [73] demonstrated synthesis of Au NPs using *Acacia nilotica* leaf extract. The phytochemicals tannins, triterpenoids, flavonoids and saponin present in the leaf extract can reduce the gold ion into nanoparticle with size ranging from 6 nm to 12 nm. The green synthesized nanoparticles showed an excellent catalytic reduction of 4 NP to 4 AP within 6 minutes under room condition. Tamuly et al., 2013 [74] prepared Au NPs using *Gymnocladusassamicus* pod extract with a size ranging from 4.5-22 nm. The formation of nanoparticles was facilitated by the phytocompounds present in the pod extract such as kaempferol, gallic acid and protocatechuic acid act as a reducing agent. The Au NPs displayed a potential catalytic activity by reducing 4-NP to 4-AP in 15 min.

Biogenic synthesis of gold nanoparticles by exploiting *Punicagranatum* juice was reported by Dash and Bag, 2014 [75]. The extract was bound with various kinds of phytochemicals, such as polyphenols, fatty acids, terpenoids, alkaloids, sterols, aromatic compounds, amino acids and tocopherols acted as capping, reducing and stabilizing agent for the synthesis of Au NPs. The synthesized nanoparticle is of triangular, pentagonal, hexagonal shape with size ranging from 25-36 nm. The prepared nanoparticles showed a complete reduction of 4 NP to 4 AP within 12 min. Paul et al.,2014 [76] synthesized biogenic Au NPs using coconut shell extract and explored their catalytic activity towards 4 NP conversation into 4 AP with the presence of NaBH$_4$. The synthesized nanoparticle size was controlled by altering the concentration of the shell extract. The reduction of 4-nitrophenol was completed in 3.0 min by using freshly prepared colloidal AuNPs. Dash et al., 2014 [77] synthesized Au NPs by exploiting polyphenolic compounds present in the extract of the *Saracaindica* barksextract at room temperature. The polyphenolic substance served as both stabilizing as well as a reducing agent. The synthesized Au NPs were of different shapes like tetragonal, pentagonal, triangular, hexagonal, and spherical with size ranging from 15–23 nm size. The catalytic efficiency of the prepared nanoparticle for 4-NP to 4-AP conversion was completed within 9 minutes at room temperature ensured that the green synthesized nanoparticles having potent catalytic efficiency towards the degradation of nitro compounds.

Choi et al., 2014[78] reported catechin mediated green synthesis of various-shaped gold nanoparticles. The catalytic property of the Au NPs was observed within 6 min in the reduction of 4-NP to 4-AP. Shape tailored gold nanoparticles were synthesized using *Garcinia Combogia* aqueous fruit extract by Rajan et al., 2014 [79]. The shape of green synthesized Au nanoparticle is of triangular, rod, hexagonal and spherical range from 12 to 22 nm in size. The catalytic activity of the green synthesized gold nanoparticles was size dependent and the reduction of 4-NP to 4-AP using sodium borohydride was completed in 20 min. A novel and straightforward and narrow sized Au NPs were prepared by using *konjacglucomannan* [80]. The polysaccharide presents in the glucomannan extract acts as both a reducing agent and stabilizer. Furthermore, Au NPs showed potent catalytic efficiency towards the reduction of 4-NP to 4-AP within 8.6 min.

Furthermore, Dash et al., 2015 [81] prepared Au NPs by using the leaf extract of *Lantana camara* Linn. The leaf extract contains different secondary metabolites like triterpenoids, flavonoids which served as reducing and capping agent. The green synthesized Au NPs were spherical shaped with 7nm in size. The synthesized Au NPs showed eminent activity towards the catalytic conversion of 4-NP to 4-AP within 7 min in the presence of NaBH$_4$. Anand et al., 2015 [82] demonstrated the catalytic efficiency of Au nanoparticles prepared via green synthesis using *Moringaoleifera* flower extract. The prepared particles are triangle and hexagonal with a size distribution range from 3-5 nm. The prepared particles showed efficient catalytic activity towards the conversion of 4 NP to 4 AP within 3 min.

Bogireddy et al., 2015 reported the biogenic synthesis of Au NPs using *Sterculiaacuminata* fruit extract and verified their efficiency to degrading various organic dyes. [83]. Various polyphenolic compounds existing in the extract served as both reducing and stabilizing agents. The prepared Au NPs showed good catalytic degradation for 4-NP in 36 min, methylene blue in 12 min, methyl orange in 12 min and direct blue in 18 min.

Wu et al., 2015 [84] demonstrated an environmental-friendly and straightforward approach for the Au NPs synthesis by utilizing aspartame as a reducing agent. The FTIR results showed that aspartame played a pivotal role in Au NPs synthesis and exhibited good catalytic activity in nitrophenol reduction in 10 min. The bark extract of *Abromaaugusta* was utilized for the biogenic synthesis of Au NPs [85]. The bark extract contains polyphenols, which served as both stabilizing and reducing agent. The Au NPs demonstrated catalytic activity in presence of sodium borohydride for the catalytic reduction of 4-NP to 4-AP in 7.5 min.

Furthermore, Yu et al., 2016 [86] synthesized biogenic Au NPs using Citrus maxima fruit extracts with particle size of about 25 nm. From the FTIR studies, phytochemicals such as polyphenol and proteins served as reducing and stabilizing agent for the formation of Au NPs. The green synthesized Au NPs showed noble catalytic efficiency towards the reduction of 4 NP to 4AP within 20 min with the excess addition of $NaBH_4$. Majumdar et al., 2016 [87] reported Biogenic formation of Au NPs from *Mimusopselengi* bark extract and studied its catalytic activity towards the conversion of 3 Nitrophenol and 4 Nitrophenol. The phytoconstituents present in the bark extract such as triterpenoids, flavonoids, tannins, polyphenols and saponin served as capping, stabilizing and reducing agents. The prepared Au NPs showed potential catalytic reduction of 4-Nitrophenol in 8 min and 3-Nitrophenol in 5 min. Ramakrishna et al., 2016 synthesized spherical shaped Au NPs with an average size of 30 nm using marine brown algae extract [88]. The aqueous marine algal extracts comprise of polysaccharides act as reducing agent. The reaction kinetics of AuNPs showed potential catalytic activity in the degradation of dyes such as Rhodamine B (10min), Sulforhodamine 101 (11 min) and 4 Nitrophenol (5min).

Banerjee and Rai, 2016 [89] studied the catalytic efficiency of green synthesized Au NPs using fungus *Aspergillus fischeri* and their potential applications towards MB dye degradation. The synthesized nanoparticles were spherical with 50 nm in size and exhibited catalytic activity for the degradation of MB dye in 9 min. Whereas, Park et al., 2016 [90] prepared gold nanoparticles using gallic acid and reported for the catalytic reduction of 4-Nitrophenol. The prepared nanoparticles exhibited potential catalytic activity towards the reduction of 4 NP to 4 AP within 8 min.

Zhang et al., 2016 [91] studied the Au NPs prepared using yeast cell *Magnusiomycesingens* extract the prepared Au particles showed various shapes including triangle, sphere and hexagon with size range from10 to 80 nm. From the FTIR studies, protein-containing amide and carboxyl groups are responsible for the gold nanoparticle formation. The prepared AuNPs showed promising catalytic activities for Nitrophenol reduction in 5 min. Assefa et al., 2017 [92] reported the microwave-assisted biogenic synthesis of gold nanoparticle using Olibanum gum extract and explored their catalytic efficacy towards the reduction of 4NP. Microwave irradiation will facilitate the faster formation of Au NPs along with the gum extract with an average size of 3 nm to 8 nm. The size of the Au NPs depends on concertation of gum extract utilized in the experiment. The as-prepared nanoparticles showed excellent catalytic activity towards potassium hexacyanoferrate (III) to potassium hexacyanoferrate (II) in 18 min and 4-NP to 4 AP in 12 min in the presence of sodium borohydride.

Sunkari et al., 2017 validated a simple, low-cost microwave mediated green synthesis of gold nanoparticles using papaya leaf extract [93]. The phytochemicals such as proteins and hydroxyl groups of alcohol and phenolic compounds were involved in the reduction of Au ion into nanoparticles. The prepared particle is about 7 to 26 nm in size and showed excellent catalytic activity towards 4 NP reduction within 11 min.

Gold nanoparticles were prepared in a size controlled manner using *Coffeaarabica* seed extract, and 4-Nitrophenol reduction was evaluated by Bogireddy et al., 2018 [94]. A well crystalline AuNPs of varied sizes were prepared by controlling the pH of the solution. From the FTIR studies, phenolic groups existing in the seed were responsible for Au NPs formation. The seed extract mediated synthesized Au NPs showed a faster catalytic reduction of 4-Nitrophenol NP to 4-AP in 16 min. A facile root of extract of *Dalbergiacoromandeliana* for Au NPs synthesis was reported by Umamaheswari et al., 2018 [95]. The chemical compound Dalspinin, isolated from the roots helped in the formation of Au NPs. The as-prepared particles showed excellent catalytic degradation against Congo red dye in 10 minutes and methyl orange in 8 minutes.

Furthermore, Gangapuram et al., 2018[96] studied the effect of microwave irradiation along with *Annona squamosa* L peel extract for the gold nanoparticle synthesis. The prepared particles showed good catalytic degradation of Congo red in 2 min, Methylene Blue in 3 min and 4-Nitrophenol in 6 min. The above results reveal that the biogenic synthesis of Au NPs using various plant materials can be used as a potential catalyst in the reduction and degradation of various toxic organic dyes for sustainable applications.

4.   Bioinspired copper nanoparticles in catalysis

Issaabadi et al., 2017 [97] reported ecofriendly biogenic copper nanoparticle (Cu NPs) synthesis using Thymus *vulgaris* L. leaf extract as both reducing agent and stabilizer without adding any kind of surfactants. The synthesized Cu NPs supported on bentonite exhibited excellent catalytic performance against toxic dyes like Congo red and methylene blue using sodium borohydride as hydrogen source at room temperature with convenient reusability and stability. Saran et al., 2018 [98] synthesized biogenic Cu NPs using *Walsuratrifoliata* flower extract anchored over TiO_2. The spherical shaped Cu NPs with average particle size of 73 nm effectively degraded the 4- Nitrophenol in 55s, Methyl orange in 5 s and Rhodamine in 6s. Ismail et al., 2018 [99] studied plant-mediated copper nanoparticle synthesis by utilizing the fruit extract of *Durantaerecta*. The as-prepared copper nanoparticles showed excellent catalytic reduction for methyl red and Congo red under the presence of sodium borohydride following pseudo-first-order reaction. Devi and Singh, 2014 [100] demonstrated the bioinspired synthesis of copper nanoparticles using *Centellaasiatica* in room temperature. With an average diameter of 2μm to 5μm, the nanoparticles exhibited catalytic degradation of methyl orange in an aqueous medium.

Kalpana et al., 2016 [101] prepared the biogenic copper nanoparticle synthesis of using the leaf extract of with particle size of about 71 nm manifesting promising catalytic activity of 89 % against Bismarck brown dye within 72 hours. Manjari et al., 2017[102] synthesized copper nanoparticles using *Aglaia elaeagnoidea* flower extract and explored it catalytic efficacy towards various organic pollutant reduction such as MB, CR and 4NP. The green synthesized copper oxide (CuO) NPs was about 45 nm in size and showed potential catalytic activity towards the catalytic conversion of MB in 10s, CR in the 30s and 4-NP in 7 min. Bio fabrication of CuONPs was synthesized using *Rheum palmatum* rootextract [103]. The as-synthesized Cu NPs showed spherical morphology with size ranging from 10 to 20 nm. From the FTIR studies, the phytochemicals such as polyphenol and flavonoids play a crucial role in the formation of NPs. The synthesized nanoparticles showed excellent catalytic activity for various organic compound reduction reactions. Curcumin conjugated green synthesis of CuOnano-grains was reported by Qasem et al., 2020 [104] and evaluated its catalytic efficiency towards methylene blue dye reduction reaction. The prepared catalyst showed a good catalytic reduction of MB dye (27 min) in the existence of $NaBH_4$ at room temperature.

5.   Biomediated synthesis of palladium nanoparticles and their catalytic applications

Kora and Rastogi, 2018 [105] studied the greenroute for palladium nanoparticles (Pd NPS) synthesis using *Anogeissuslatifolia* as reducing agent. The synthesized spherical particles were about 4.8 ± 1.6 nm in size. The as-synthesized biogenic pallidum nanoparticles showed promising catalytic activity with sodium borohydride presence against various dyes. They also synthesized palladium nanoparticles via a green route from gum olibanum (*Boswelliaserrata*), glucuronoarabinogalactan polymer with average particle size about 6.6 ± 1.5 nm. The hydroxyl, protein and carbonate functional groups present in the polymer mediated nanoparticle synthesis by behaving as capping and reducing agent. They also exhibited a reduction of dyes such as Rhodamine B, methylene blue, coomassie brilliant blue and 4-nitrophenol with sodium borohydride [106]. Biogenic palladium nanoparticles were synthesized from carboxymethylcellulose yielding particles size of about 2.5 nm. The as-synthesized palladium nanoparticles were found to be spherical with a negative zeta potential of -52.6 mV. They exhibited promising catalytic degradation efficiency with sodium borohydride for the dyes such as acid orange 7, p- aminobenzene, scarlet 3G, acid red 66 and reactive yellow 179 [107]. Bordbar and Mortazavimanesh, 2017 [108] immobilized palladium nanoparticles on the walnut shell powder surface using the leaf extract of *Equisetum arvense* L. to examine the catalytic activity of various organic dyes. Palladium/ walnut shell composite showed higher catalytic activity with methylene blue, 4- Nitrophenol, Congo red and Rhodamine B with sodium borohydride at room temperature. This low cost, alterable supported composite also exhibited recycling and recovery without any appreciable loss up to seven times. Garai et al., 2018 [109] prepared biogenic palladium nanoparticles from the extract of *Terminalia arjuna bark* with a diameter size of 8.9 nm and examined the efficiency for Suzuki, Heck type C-C coupling reaction and catalytic degradation of dyes. Suzuki reaction was carried out in the presence of biogenic palladium nanoparticles using iodobenzene, phenylboronic acid in the presence of water at 100 ° C. The reaction yielded 99.3 % purified product with TON and TOF as 1241 and 138 h^{-1} respectively. Whereas, the Heck reaction yielded 99.5 % purified product with TON and TOF as 996 and 190 h^{-1}. The synthesized Pd colloidal nanoparticles showed promising catalytic activity towards Rhodamine B and methylene blue with sodium borohydride. Nasrollahzadeh et al., 2015 [110] prepared biogenic palladium nanoparticles using the leaf extract of *Hippophaerhamnoides* as both capping and reducing agent with particle size ranging between 2.5 nm and 14 nm. The as-prepared heterogeneous catalyst yielded 100 % of the yield in Suzuki–Miyaura coupling reaction under lignin-free conditions in water. Whereas the different aryl halides induced Suzuki–Miyaura coupling reaction produced only 91-95 % of the yield. Nasrollahzadeh et al., 2016 [111] fabricated biogenic palladium nanoparticles supported onto reduced graphene oxide using bearberry fruit and examined the catalytic reduction of nitroarenes to amines in the water-alcohol mixture. Palladium nanoparticles reduced the nitrobenzene to aniline in the presence of sodium borohydride. The nitrobenzene reduction takes place in a stepwise manner by adsorption of H_2-, electron transfer from sodium borohydride and the desorption of amino compounds from the catalyst surface. The prepared catalyst exhibited catalytic activity without any significant loss up to five cycles. Khan et al., 2017 [112] prepared bioinspired palladium nanoparticles from the aqueous extract of *Pulicariaglutinosa* acting as both bioreductant and capping ligand. The as-synthesized palladium nanoparticles showed promising catalytic activity for the Suzuki coupling reaction with kinetic studies manifesting the completion of reaction within five minutes. Similarly, Nasrollahzadeh and Sajadi, 2016 [113] reported excellent catalytic property of biogenic palladium nanoparticles prepared from the leaf extract of E*uphorbia granulate* for the phosphine free Suzuki–Miyaura coupling reactions of different aryl halides. Khan et al., 2017 [114] also reported the green synthesis of Pd NPs using *Salvadorapersica* L. root extract for the superior catalytic activity for the Suzuki coupling reactions of different aryl halides in the aqueous medium. The as-synthesized palladium also exhibited promising reusability, and the kinetic studies revealed that the conversion of various aryl halides into biphenyl within a short period.

6.　Bio-fabrication of platinum nanoparticles and their catalytic applications

The use of biogenic compounds for the synthesis of particles is of enormous interest to modern nanotechnology. Venu et al., 2011 [115] developed a novel and biogenic method for the preparation of platinum nanoparticles utilizing honey. From the FTIR analysis proteins and polyphenols play a pivotal role in platinum nanoparticles formation. The prepared Pt NPs showed excellent catalytic conversion of aniline with 4-amino antipyrine to Antipyrilquinoneimine dye in an acidic aqueous medium was achieved in 70 s.

Likewise, Sheny et al., 2013 [116] synthesized platinum nanoparticles using the dried leaf extract of *Anacardiumoccidentale*. The prepared Pt NPs particles showed irregular rod-shaped with a crystalline structure and efficiently reduced 4 NP to 4 AP within 30 minutes. From the FTIR analysis, the protein present in the leaf extract facilitates the formation of Pt NPs by acting as reducing and stabilizing agent.

Similarly, Pandy et al., 2014 [117] reported the novel, nontoxic, biogenic synthesis of Pt nanoparticles using guar gum. The prepared particles were utilized for the liquid phase reduction of *p*-Nitrophenol (*p*-NP) into *p*-aminophenol (*p*-AP). The efficacy of Pt NPs in catalytic reduction of *p*-NP was reported to be 97% in a total time of 5.2 min.

Biofabrication of Pt NPs was reported by Dauthal et al., 2015 [118] using *Punicagranatum* peel extract. The prepared particles were spherical with a size ranging from 16 to 23 nm. The hydroxyl group of phenolic moiety present in peel extract acted as a reducing agent during the synthesis of Pt NPs. The catalytic reduction of 3-nitrophenol was completed in less than six minutes in the presence of NaBH$_4$.

Ye et al.,2016 [119] fabricated dendrimer-like Pt alloy nanoparticles supported on the surface of PDA/RGO using ascorbic acid. From the TEM analysis dendrimer nanostructures was formed by assembling of a cluster of spherical Pt NPs with the size of about 2.7 nm. The as-prepared PtPDA/RGO showed promising catalytic activity by reducing 4- NP to AP within 12 min.

Furthermore, Ramkumar et al., 2017 [120] described a one-step synthesis of Pt NPs using Indian brown seaweed *Padinagymnospora* and their catalytic activity. From the FTIR spectrum, it is inferred that phytocompounds like carbohydrates and proteins present in them reduced platinum ions (Pt+) into Pt NPs. The resultant Pt NPs were truncated octahedral in shape with size from 5–50 nm.

A facile bioinspired route for the preparation of Pt NPs using *Fumariaeherba* extract was reported by Dobrucka 2019 [121]. The TEM results reveal that the prepared particles are hexagonal and pentagonal, with a size of about 30 nm. The phytochemicals such as alkaloids, flavonoid compounds and phenolic acid present in the plant extract mediated the formation of Pt NPs. Moreover, platinum nanoparticles presented promising catalytic properties by reducing methylene blue in 15 min and crystal violet dye in 15 min. Figure 3 displayed the catalytic reduction of various dyes using green synthesized metal nanoparticles.

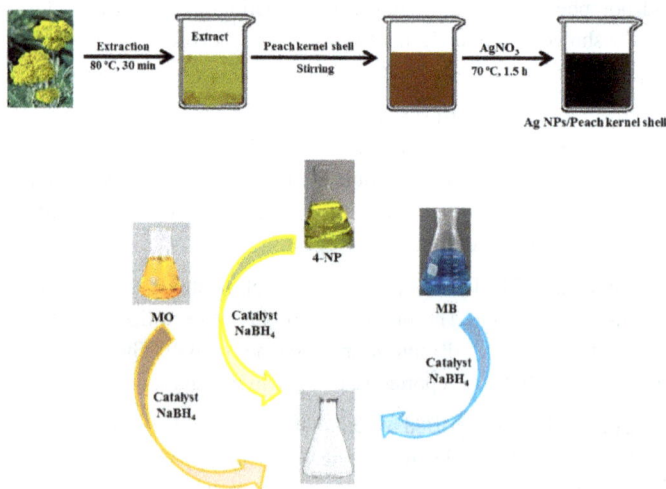

Figure.7. Catalytic reduction of various dyes using green synthesized nanoparticles. Reproduced permission from Ref. No. [103]

7.   Green synthesis of iron nanoparticles and their catalytic applications

Hoag et al., 2009 [122] investigated the degradation of bromothymol blue by catalyzing hydrogen peroxide with biogenic iron nanoparticles prepared from green tea leaves. The catalytic activity was more in green tea synthesized nanoparticles than of iron nanoparticles prepared from EDT A and EDDS complexes. Similarly, bioinspired iron nanoparticles produced from the Sorghum bran extract efficiently degraded bromomethyl blue. In the presence of hydrogen peroxide, iron nanoparticles catalyze the free radical production from hydrogen peroxide by increasing the rate of degradation [123]. Shahwan et al., 2011 [124] achieved complete degradation of methyl orange and methyl blue dyes with green tea leaves mediated iron nanoparticles at 350 minutes and 200 minutes, respectively. Also, the green tea mediated iron nanoparticles were proved to be more competent in terms of percentage removal and kinetics when compared with borohydride reduced iron nanoparticles. Huang et al., 2013 [125] examined the degrading efficiency of the green synthesized iron nanoparticles from oolong tea extract. The oolong tea extract acted both as capping and reducing agent, leading to increased reactivity and reduced aggregation of iron nanoparticles. The degradation efficiency was 75.5% of Malachite Green with a concentration of 50 mg/L.

Similarly, three types of tea extracts, namely black tea, green tea, oolong tea were used for the synthesis of iron nanoparticles and tested for catalytic degradation of monochlorobenzene. The degradation rate was higher in green tea inspired iron nanoparticles owing to the high phenol content [126]. Huang et al., 2015 [127] further ventured into the experimental aspects like Fe^{2+} volume ratio and tea extract, pH, temperature to study their influence on nanoparticles synthesis. It was found that iron nanoparticle concentration decreased with an increase in leaf extract concentration due to declining Fe^{2+} concentration. The catalytic degradation of Malachite Green was found to be influenced by synthesis condition and pH, out of which temperature influencing the reactivity. Luo et al., 2014 [128] manifested the bioinspired iron nanoparticles from the grape leaf extract as Fenton like catalyst with hydrogen peroxide. It was found that they were promisingly effective in catalytic degradation of different types of organic dyes in different experimental conditions in a sustainable manner. Truskewycz et al., 2016 [129] also demonstrated that amorphous iron nanoparticles derived from green tea extract could catabolize anthraquinone and other azo dye mixtures up to 90 % within 20 minutes at 70 °C with a concentration of 0.53 ppm. The experimental evidence proved that catalytic activity is driven by temperature and pH conditions favoring temperature between 70 and 90 °C and lower pH. A detailed comparison of green synthesis of various metal nanoparticles using different plant part extracts and their catalytic applications was described in Table 1.

Table 1. Comparison of various green synthesized metal nanoparticles using different plant extracts and their catalytic applications.

S.L. No	Plant or biomaterial name	Part utilized	Metal NPs	Size of NPs (nm)	Catalytic application	Reduction Time (min)	Ref.
1.	*Mussaendaglabrata*	Leaf	Ag	51.3	4 Nitrophenol	9	45
					Rhodamine B	7	
					Methyl orange	4	
2.	*Biophytumsensitivum*	Leaf	Ag	19.6	Methylene blue	9	46
					Methyl orange	11	
3.	*Zanthoxylumarmatum*	Leaf	Ag	35	Methyl Orange	1440	47
					Methylene Blue	1440	
					Methyl Red	1440	
					Saffranin O	1440	
4.	*Gmelinaarborea*	Fruit	Ag	17	Methylene Blue	15	48
5.	*Amarranthusgangeticus*	Leaf	Ag	13	Congo red	15	49
6.	*foenum-graecum*	Seed	Ag	28	Methylene blue,	8	50
					Methyl orange	6	
					Eosin y	10	
7.	*Plumeria alba*	Flower	Ag	36	4 Nitrophenol	8	51
					Methylene blue	120	
					Ethidium bromide	180	
8.	*Thymbraspicata*	Leaf	Ag	7	Rhodamine B	1	52
					4- Nitrophenol	1	
					Methylene blue	50s	
9.	Mussaendaerythrophylla	Leaf	Ag	88	Methyl orange	45	53
10.	Grape	Seed	Ag	54	Direct Orange 26	18	54
11.	Tea	Waste tea	Ag	45	Methylene blue	20	56
12.	*Elephantopusscaber*	Leaf	Ag	37	4-Nitrophenol	7	57
					2-Nitroaniline	25	
					4-Nitroaniline	7	
					Eosin y	16	
13.	*Prosopisjuliflora*	Bark	Ag	55	4-Nitrophenol	80	58
14.	*Terminalia cuneata*	Bark	Ag	25	Direct yellow-12	40	59
15.	Desmostachyabipinnata	Leaf	Ag	10-35	Methylene blue	2	60
					Methyl red	2	
					Congo red	2	
					Acridine orange	2	
16.	*Sterculiaacuminata*	Fruit	Ag	10	4-Nitrophenol	22	61
					Methylene blue	3	
					Methyl orange	3	
					Phenol red	3	

					Direct blue 24	6	
17.	*Anacardiumoccidentale*	Testa	Ag	25	Congo red	25	62
					Methyl orange	15	
18.	*Ekebergiacapensis*	Leaf	Ag	25	Allura red	45	65
19.	*Cassia auriculata*	Flower	Ag	10 to 30	Methyl organe	30	66
					4-Nitrophenol	12	
20.	*Aglaia elaeagnoidea*	Flower	Ag	17	Methylene blue	5s	67
					Congo red	10s	
					4 Nitrophenol	15	
21.	Nervaliazeylanica	Leaf	Ag	34.2	Methyl orange	10	68
					Rhodamine B	7	
22.	*Trigonellafoenum-graecum*	seed	Au	15 to 25	4 Nitrophenol	7	69
23.	*Prunusdomestica*	Fruit	Au	20	4 Nitrophenol	20	70
24.	*Cochlospermumreligiosum*	Gum	Au	7	4 Nitrophenol	30	71
25.	*Pleurotusflorida*	Mushroom	Au	10	4 Nitrophenol	30	72
26.	*Acacia nilotica*	Leaf	Au	8	4 Nitrophenol	6	73
27.	*Gymnocladusassamicus*	Pod	Au	4.5 to 22	4 Nitrophenol	15	74
28.	*Punicagranatum*	Fruit	Au	25 to 36	4 Nitrophenol	12	75
29.	*Cocos nucifera*	Shell	Au	15	4 Nitrophenol	3	76
30.	*Saracaindica*	Bark	Au	15 to 23	4 Nitrophenol	9	77
31.	*Garcinia Combogia*	Fruit	Au	12 to 22	4 Nitrophenol	20	79
32.	*Lantana camara*	Leaf	Au	7	4 Nitrophenol	7	81
33.	*Moringaoleifera*	Flower	Au	5	4 Nitrophenol	3	82
34.	*Sterculiaacuminata*	fruit	Au		4-nitrophenol	36	83
					Methylene blue	12	
					Methyl orange	12	
					Direct blue 24	18	
35.	*Abromaaugusta*	bark	Au	33	4 Nitrophenol	7.5	85
36.	*Citrus maxima*	fruit	Au	25	4 Nitrophenol	20	86
37.	*Mimusopselengi*	bark	Au	9 to 14	4- Nitrophenol	8	87
					3-Nitrophenol	5	
38.	*Olibanum gum*	Gum	Au	3 to 8	Hexacyanoferrate	18	92
					4- Nitrophenol	12	
39.	*Carica papaya*	Leaf	Au	7 to 26	4 Nitrophenol	11	93
40.	*Coffeaarabica*	seed	Au	12	4 Nitrophenol	16	94
41.	*Dalbergiacoromandeliana*	root	Au	11	Congo rcd	10	95
					methyl orange	8	
42.	*Annona squamosa*	Fruit	Au	5	Congo red	2	96
					Methylene Blue	3	
					4- Nitrophenol	6	

43.	*Thymus vulgaris*	Leaf	Cu	56	Methylene blue	0.4	97
					Congo red	5	
44.	*Walsuratrifoliata*	Flower	CuO	73	4- Nitrophenol	1	98
					Methyl orange	5s	
					Rhodamine B	6s	
45.	*Durantaerecta*	Fruit	Cu	70	Methyl orange	4	99
					Congo red	5	
46.	*Centellaasiatica*	Leaf	CuO	200	Methyl orange	24 hrs	100
47.	*Tridaxprocumbens*	Leaf	Cu	71	Bismarck brown	72 hrs	101
48.	*Aglaia elaeagnoidea*	Flower	CuO	45	Methylene blue	10s	102
					Congo red	30s	
					4- Nitrophenol	7	
49.	*Rheum palmatum*	Root	CuO	10 - 20	4- Nitrophenol	22	103
					Methyl blue	8	
					Rhodamine B	10	
50.	*Curcumin*	Curcumin	Cu	-	Methylene blue	27	104
51.	*Anogeissuslatifolia*	Gum	Pd	4.8	Methylene Blue	2	105
					Coomassie Blue	2	
					Methyl Orange	2	
					4- Nitrophenol	2	
52.	*Boswelliaserrata*	Gum	Pd	6.6	Methylene Blue	2	106
					Coomassie Blue	2	
					Methyl Orange	2	
					4- Nitrophenol	2	
53.	*carboxymethyl cellulose*	carboxymethyl cellulose	Pd	2.5	p- Aminoazobenzene	60	107
					Acid red 66	60	
					Acid orange 7	1.5	
54.	*Equisetum arvense*	Leaf	Pd	5–12	4- Nitrophenol	1	108
					Methyl Orange	1s	
					Methylene Blue	1s	
					Congo red	25s	
					Rhodamine B	13s	
55.	*Terminalia arjuna*	Bark	Pd	9	Methylene Blue	9	109
					Rhodamine B	9	
56.	*Hippophaerhamnoides*	Leaf	Pd	5	Suzuki–Miyaura reaction	12 hrs	110
57.	*Berberis vulgaris*	Fruit	Pd	18	4- Nitrophenol	1.5 hrs	111
58.	*Pulicariaglutinosa*	Leaf	Pd	25	Suzuki–Miyaura reaction	4	112
59.	*Euphorbia granulate*	Leaf	Pd	27	Suzuki–Miyaura reaction	3 hrs	113
60.	*Salvadorapersica*	root	Pd	5	Suzuki–Miyaura reaction	3	114
61.	*Anacardiumoccidentale*	Leaf	Pt	500	4- Nitrophenol	30	116
62.	Guar gum	Gum	Pt	6	4- Nitrophenol	5.2	117

63.	*Punicagranatum*	Fruit Peel	Pt	16 to 23	3- Nitrophenol	6	118
64.	Ascorbic Acid	Ascorbic acid	Pt	2.7	4- Nitrophenol	12	119
65.	*Fumariaeherba*	Leaf	Pt	30	Methylene blue	15	121
					Crystal violet	15	
66.	*Camellia sinensis*	Leaf	Fe	5 to 15	Bromothymol blue	10	122
67.	*Sorghum spp*	Bran	Fe	50	Bromothymol blue	30	123
68.	Green tea	Leaf	Fe	40	Methylene blue	120	124
					Methyl Orange	120	
69.	Oolong tea	Leaf	Fe	40 to 50	Malachite green	40	125
70.	*Camellia sinensis*	Leaf	Fe	20 to 40	monochlorobenzene	180	126
71.	Green tea	Leaf	Fe	70 to 80	Malachite green	50	127
72.	Grape	Leaf	Fe	60	Acid Orange	120	128
73.	Green tea	Leaf	Fe	20 to 50	Brilliant Blue	20	129
					Direct Red	20	

8. Mechanism of catalytic conversion of 4- Nitrophenol to 4- Aminophenol using green synthesized metal nanoparticles

The catalytic potential of the biogenic metal nanoparticles was explored for the catalytic hydrogenation of 4-NP into 4-AP in excess of sodium borohydride solution. This reaction is thermodynamically feasible, but kinetically not favorable due to long potential difference between the acceptor and donor molecule. Addition of metal nanoparticles initiates the catalytic reaction by enabling the transfer of electron from BH4– donor to 4-NP acceptor. The reduction of 4 NP to 4- AP happens with the formation of intermediate 4- Nitrophenolate ion. This can be explained by the shift of absorbance peak from 317 nm to 400nm after the addition of NaBH4 owing to the formation of 4- nitrophenolate ion. The reaction progress has been evaluated by tracking the alterations in the 4- nitrophenolate absorption spectra at 400 nm. After the addition of metal nanoparticles, the absorption peak intensity at 400 nm reduces quickly, whereas a parallel new peak appeared at 298 nm indicates the formation of 4-Aminophenol [98, 130]. The plausible mechanism of 4-NP reduction to 4-AP using green synthesized nanoparticles was displayed in Figure 4.

Bioinspired Nanomaterials for Energy and Environmental Applications   Materials Research Forum LLC
Materials Research Foundations **121** (2022) 1-38   https://doi.org/10.21741/9781644901830-1

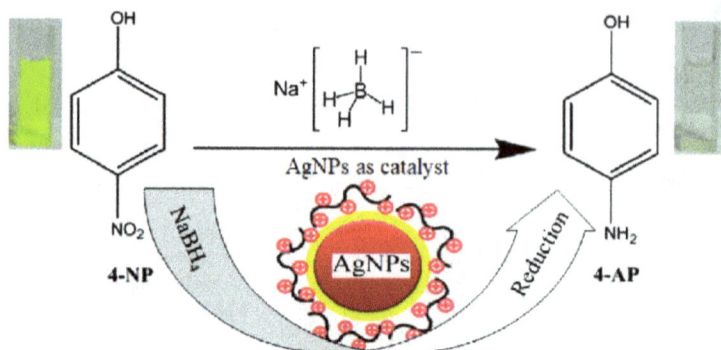

Figure.8. Mechanism of catalytic reduction of 4 Nitrophenol to 4 Aminophenol using green synthesized nanoparticles. Reproduced permission from Ref. No. [130]

Conclusion

Green synthesis of metal nanoparticles utilizing different plant parts such as leaves, flowers, bark, gum, fruit and root extract was discussed in this review. The phytochemicals and secondary metabolites in the plant extract such as terpenoids, flavonoids, saponins, carbohydrates, ketones, aldehydes, carboxylic acids, amides, proteins and vitamins cans serve as reducing, stabilizing and capping agents in the formation of metal nanoparticles. These green synthesized nanoparticles having different morphology with spherical, triangle, hexagonal and irregular in shapes with a size ranging from 2 nm to 60 nm. These biogenic synthesized nanoparticles have been successfully utilized for various catalytic applications. The prepared catalyst showed potential catalytic activity towards the reducing and degradation of various organic pollutants. Based on the above discussion, the smaller the size of the nanoparticles having higher catalytic activity. Hence it is projected that the green synthesis of metal nanoparticles from plant materials having a bright prospect in the field of catalysis and other applications.

Acknowledgement

The author Dr. Saran. S is thankful to IIT Guwahati for providing Institutional Post-Doctoral fellowship.

Conflict of interest

The authors declare that they have no conflict of interest.

References

[1] J.H. Fendler, Nanoparticles and nanostructured films: preparation, characterization, and applications, John Wiley & Sons (2008).

[2] R.M. Crooks, M. Zhao, L. Sun, V. Chechik, L.K. Yeung, Dendrimer-encapsulated metal nanoparticles: synthesis, characterization, and applications to catalysis, Acc. Chem. Res, 34(2001) 181-190. https://doi.org/10.1021/ar000110a

[3] M.P. Ferraz, F.J. Monteiro, C.M. Manuel, Hydroxyapatite nanoparticles: a review of preparation methodologies, J. Appl.Biomater.Biomech. 2(2004) 74-80.

[4] F. Mafuné, J.Y. Kohno, Y. Takeda, T. Kondow, Full physical preparation of size-selected gold nanoparticles in solution: laser ablation and laser-induced size control, J. Phys. Chem. B, 106 (2002) 7575-7577. https://doi.org/10.1021/jp020577y

[5] A. Fujita, Y. Matsumoto, M. Takeuchi, H. Ryuto, G.H. Takaoka, Growth behavior of gold nanoparticles synthesized in unsaturated fatty acids by vacuum evaporation methods, Phys. Chem. Chem. Phys.18 (2016) 5464-5470. https://doi.org/10.1039/C5CP07323E

[6] F.E. Kruis, H. Fissan, A. Peled, Synthesis of nanoparticles in the gas phase for electronic, optical and magnetic applications—a review, J. Aerosol Sci. 29 (1998) 511-535. https://doi.org/10.1016/S0021-8502(97)10032-5

[7] A. Dhakshinamoorthy, A.M. Asiri, H. Garcia, Metal–organic frameworks catalyzed C–C and C–heteroatom coupling reactions, Chem. Soc. Rev. 44(2015) 1922-1947. https://doi.org/10.1039/C4CS00254G

[8] B.C. Gates, Supported metal clusters: synthesis, structure, and catalysis, Chem. Rev. 95 (1995) 511-522. https://doi.org/10.1021/cr00035a003

[9] R.P. Singh, S. Magesh, C. Rakkiyappan, Formation of fenugreek (Trigonellafoenum-graecum) extract mediated Ag nanoparticles: mechanism and applications, Int. J. Bioeng. Sci. Technol. 2(2011), 64-73.

[10] M. Kasithevar, M. Saravanan, P. Prakash, H. Kumar, M. Ovais, H. Barabadi, Z.K. Shinwari, Green synthesis of silver nanoparticles using Alysicarpusmonilifer leaf extract and its antibacterial activity against MRSA and CoNS isolates in HIV patients, J. Interdiscip. Nanomed. 2(2017) 131-141. https://doi.org/10.1002/jin2.26

[11] J. Singh, T. Dutta, K.H. Kim, M. Rawat, P. Samddar, P. Kumar, Green' synthesis of metals and their oxide nanoparticles: applications for environmental remediation, J.Nanobiotechnol. 16(2018) 84. https://doi.org/10.1186/s12951-018-0408-4

[12]   P. Deepak, V. Amutha, C. Kamaraj, G. Balasubramani, D. Aiswarya, P. Perumal, Chemical and green synthesis of nanoparticles and their efficacy on cancer cells. In Green Synthesis, Characterization and Applications of Nanoparticles, Elsevier (2019) 369-387. https://doi.org/10.1016/B978-0-08-102579-6.00016-2

[13]   S. Patra, S. Mukherjee, A.K. Barui, A. Ganguly, B.Sreedhar, C.R.Patra, Green synthesis, characterization of gold and silver nanoparticles and their potential application for cancer therapeutics, Mater. Sci. Eng. C 53 (2015) 298-309. https://doi.org/10.1016/j.msec.2015.04.048

[14]   T.J.I. Edison, M.G. Sethuraman, Instant green synthesis of silver nanoparticles using Terminalia chebula fruit extract and evaluation of their catalytic activity on reduction of methylene blue, Process Biochem. 47 (2012) 1351-1357. https://doi.org/10.1016/j.procbio.2012.04.025

[15]   M. Ayaz, M. Junaid, F. Ullah, F. Subhan, A. Sadiq, G. Ali, M. Ovais, M. Shahid, A. Ahmad, A. Wadood, M. El-Shazly, Anti-Alzheimer's studies on β-sitosterol isolated from Polygonumhydropiper L., Front. Pharmacol. 8 (2017) 697. https://doi.org/10.3389/fphar.2017.00697

[16]   M. Ovais, A.T. Khalil, N.U. Islam, I. Ahmad, M. Ayaz, M. Saravanan, Z.K. Shinwari, S. Mukherjee, Role of plant phytochemicals and microbial enzymes in biosynthesis of metallic nanoparticles, Appl. Microbiol. Biotechnol. 102 (2018) 6799-6814.

[17]   A.K. Jha, K. Prasad, K. Prasad, A.R. Kulkarni, Plant system: nature's nanofactory, Colloids Surf., B 73 (2009) 219-223. https://doi.org/10.1016/j.colsurfb.2009.05.018

[18]   S. Raja, V. Ramesh, V. Thivaharan, Green biosynthesis of silver nanoparticles using Calliandrahaematocephala leaf extract, their antibacterial activity and hydrogen peroxide sensing capability, Arab. J. Chem. 10 (2017) 253-261. https://doi.org/10.1016/j.arabjc.2015.06.023

[19]   J. Lee, H.Y. Kim, H. Zhou, S. Hwang, K. Koh, D.W. Han, J. Lee, Green synthesis of phytochemical-stabilized Au nanoparticles under ambient conditions and their biocompatibility and antioxidative activity, *J. Mater. Chem.* 21 (2011) 13316-13326. https://doi.org/10.1039/c1jm11592h

[20]   Y. Zhou, W. Lin, J. Huang, W. Wang, Y. Gao, L. Lin, M. Du, Biosynthesis of gold nanoparticles by foliar broths: roles of biocompounds and other attributes of the extracts, Nanoscale Res. Lett, 5 (2010) 1351. https://doi.org/10.1007/s11671-010-9652-8

[21]   N. Ahmad, S. Sharma, M.K. Alam, V.N. Singh, S.F. Shamsi, B.R. Mehta, A. Fatma, Rapid synthesis of silver nanoparticles using dried medicinal plant of basil, Colloids Surf. B 81 (2010) 81-86. https://doi.org/10.1016/j.colsurfb.2010.06.029

[22]   S.M. Ghoreishi, M. Behpour, M. Khayatkashani, Green synthesis of silver and gold nanoparticles using Rosa damascena and its primary application in electrochemistry, Physica E Low Dimens. Syst. Nanostruct.44 (2011) 97-104. https://doi.org/10.1016/j.physe.2011.07.008

[23]   N. Sahu, D. Soni, B. Chandrashekhar, D.B. Satpute, S. Saravanadevi, B.K. Sarangi, R.A. Pandey, Synthesis of silver nanoparticles using flavonoids: hesperidin, naringin and diosmin, and their antibacterial effects and cytotoxicity, Int. Nano Lett. 6 (2016) 173-181. https://doi.org/10.1007/s40089-016-0184-9

[24]   P. Trouillas, P. Marsal, D. Siri, R. Lazzaroni, J.L. Duroux, A DFT study of the reactivity of OH groups in quercetin and taxifolin antioxidants: The specificity of the 3-OH site, Food Chem. 97 (2006) 679-688. https://doi.org/10.1016/j.foodchem.2005.05.042

[25]   K. Yoosaf, B.I. Ipe, C.H. Suresh, K.G. Thomas, In situ synthesis of metal nanoparticles and selective naked-eye detection of lead ions from aqueous media, J. Phys. Chem. C, 111 (2007) 12839-12847. https://doi.org/10.1021/jp073923q

[26]   A.E. Edris, Pharmaceutical and therapeutic potentials of essential oils and their individual volatile constituents: a review. Phytotherapy Research: An International Journal Devoted to Pharmacological and Toxicological Evaluation of Natural Product Derivatives, 21(2007) 308-323. https://doi.org/10.1002/ptr.2072

[27]   JYBS Kim, Rapid biological synthesis of silver nanoparticles using plant leaf extracts, Bioprocess Biosyst. Eng. 32 (2009) 79. https://doi.org/10.1007/s00449-008-0224-6

[28]   D.S. Sheny, J. Mathew, D. Philip, Synthesis characterization and catalytic action of hexagonal gold nanoparticles using essential oils extracted from Anacardiumoccidentale, Spectrochim. Acta Part A, 97 (2012) 306-310. https://doi.org/10.1016/j.saa.2012.06.009

[29]   EC Da Silva, M.G.A. Da Silva, S.M.P. Meneghetti, G. Machado, M. A, R. C. Alencar, J. M. Hickmann, M.R. Meneghetti, Synthesis of colloids based on gold nanoparticles dispersed in castor oil, J.Nanopart. Res. 10 (2008) 201-208. https://doi.org/10.1007/s11051-008-9483-z

[30] V. Vilas, D. Philip, J. Mathew, Catalytically and biologically active silver nanoparticles synthesized using essential oil,Spectrochim. Acta Part A 132 (2014) 743-750. https://doi.org/10.1016/j.saa.2014.05.046

[31] M. M. Kumari, D.Philip, Facile one-pot synthesis of gold and silver nanocatalysts using edible coconut oil, Spectrochim. Acta, Part A, 111 (2013). 154-160. https://doi.org/10.1016/j.saa.2013.03.076

[32] A.K. Singh, M. Talat, D.P. Singh, O.N. Srivastava, Biosynthesis of gold and silver nanoparticles by natural precursor clove and their functionalization with amine group, J. Nanopart. Res. 12 (2010) 1667-1675. https://doi.org/10.1007/s11051-009-9835-3

[33] S. Li, Y. Shen, A. Xie, X. Yu, L. Qiu, L. Zhang, Q. Zhang. Green synthesis of silver nanoparticles using Capsicum annuum L. extract, Green Chem. 9 (2007) 852-858. https://doi.org/10.1039/b615357g

[34] S. Mukherjee, D. Chowdhury, R. Kotcherlakota, S. Patra, Potential theranostics application of bio-synthesized silver nanoparticles (4-in-1 system), Theranostics4 (2014) 316. https://doi.org/10.7150/thno.7819

[35] R. Shukla, S.K. Nune, N. Chanda, K. Katti, S. Mekapothula, RR. Kulkarni, WV. Welshons, R. Kannan, K.V. Katti, Soybeans as a phytochemical reservoir for the production and stabilization of biocompatible gold nanoparticles, Small 4 (2008)1425-1436. https://doi.org/10.1002/smll.200800525

[36] P. Raveendran, J. Fu, SL. Wallen, Completely "green" synthesis and stabilization of metal nanoparticles, J. Am. Chem. Soc. 125 (2003) 13940-13941. https://doi.org/10.1021/ja029267j

[37] X. Zhao, Y. Xia, Q. Li, X. Ma, F. Quan, C. Geng, Z. Han, Microwave-assisted synthesis of silver nanoparticles using sodium alginate and their antibacterial activity, Colloids Surf. A 444 (2014) 180-188. https://doi.org/10.1016/j.colsurfa.2013.12.008

[38] AJ González Fá, A. Juan, MS Di Nezio, Synthesis and characterization of silver nanoparticles prepared with honey: the role of carbohydrates, Anal. Lett. 50 (2017) 877-888. https://doi.org/10.1080/00032719.2016.1199558

[39] M. Shah, D. Fawcett, S. Sharma, SK. Tripathy, G.E.J. Poinern, Green synthesis of metallic nanoparticles via biological entities, Materials 8 (2015) 7278-7308. https://doi.org/10.3390/ma8115377

[40]   K.S. Chou, C.Y. Ren, Synthesis of nanosized silver particles by chemical reduction method, Mater Chem Phys. 64 (2000) 241–246. https://doi.org/10.1016/S0254-0584(00)00223-6

[41]   G. Guzman, J. Dille, S. Godet, Synthesis of silver nanoparticles by chemical reduction method and their antibacterial activity, Int. J. Chem. Biol. Eng. 2 (2009) 104–111.

[42]   Q.H. Tran, A.T. Le, Silver nanoparticles: synthesis, properties, toxicology, applications and perspectives, Adv. Nat. Sci. Nanosci. Nanotechnol. 4 (2013) 033001. https://doi.org/10.1088/2043-6262/4/3/033001

[43]   G. Reddy, J. Joy, T. Mitra, S. Shabnam, T. Shilpa, Nanosilver review, Int. J. Adv. Pharm. 2 (2012) 9–15.

[44]   S. Ahmed, M. Ahmad, B.L. Swami, S. Ikram, A review on plants extract mediated synthesis of silver nanoparticles for antimicrobial applications: a green expertise, J. Adv. Res. 7 (2016) 17–28. https://doi.org/10.1016/j.jare.2015.02.007

[45]   S. Francis, S. Joseph, E.P. Koshy, B. Mathew, Green synthesis and characterization of gold and silver nanoparticles using Mussaendaglabrata leaf extract and their environmental applications to dye degradation, Environ. Sci. Poll. Res. 24(2017) 17347-17357. https://doi.org/10.1007/s11356-017-9329-2

[46]   S. Joseph, B. Mathew, Microwave-assisted green synthesis of silver nanoparticles and the study on catalytic activity in the degradation of dyes, J. Mol. Liq. 204 (2015) 184-191. https://doi.org/10.1016/j.molliq.2015.01.027

[47]   K. Jyoti, A. Singh, Green synthesis of nanostructured silver particles and their catalytic application in dye degradation, J. Genet. Eng. Biotechnol. 14(2016) 311-317. https://doi.org/10.1016/j.jgeb.2016.09.005

[48]   J. Saha, A. Begum, A. Mukherjee, S. Kumar, A novel green synthesis of silver nanoparticles and their catalytic action in reduction of Methylene Blue dye, Sustain. Environ. Res. 27 (2017) 245-250. https://doi.org/10.1016/j.serj.2017.04.003

[49]   H. Kolya, P. Maiti, A. Pandey, T. Tripathy, Green synthesis of silver nanoparticles with antimicrobial and azo dye (Congo red) degradation properties using Amaranthusgangeticus Linn leaf extract, J. Anal. Sci. Technol. 6 (2015) 33. https://doi.org/10.1186/s40543-015-0074-1

[50]   V.K. Vidhu, D. Philip, Catalytic degradation of organic dyes using biosynthesized silver nanoparticles, Micron 56 (2014) 54-62. https://doi.org/10.1016/j.micron.2013.10.006

[51]　R. Mata, J.R. Nakkala, S.R. Sadras, Catalytic and biological activities of green silver nanoparticles synthesized from Plumeria alba (frangipani) flower extract, Mater. Sci. Eng. C 51 (2015) 216-225. https://doi.org/10.1016/j.msec.2015.02.053

[52]　H. Veisi, S. Azizi, P. Mohammadi, Green synthesis of the silver nanoparticles mediated by Thymbraspicata extract and its application as a heterogeneous and recyclable nanocatalyst for catalytic reduction of a variety of dyes in water, J. Clean. Prod. 170 (2018) 1536-1543. https://doi.org/10.1016/j.jclepro.2017.09.265

[53]　T. Varadavenkatesan, R. Selvaraj, R. Vinayagam, Phyto-synthesis of silver nanoparticles from Mussaendaerythrophylla leaf extract and their application in catalytic degradation of methyl orange dye, J. Mol. Liq. 221 (2016) 1063-1070. https://doi.org/10.1016/j.molliq.2016.06.064

[54]　Y. Ping, J. Zhang, T. Xing, G. Chen, R. Tao, K.H. Choo, Green synthesis of silver nanoparticles using grape seed extract and their application for reductive catalysis of Direct Orange 26, Ind. Eng. Chem. Res. 58 (2018) 74-79. https://doi.org/10.1016/j.jiec.2017.09.009

[55]　Y. Junejo, A. Baykal, M. Safdar, A. Balouch, A novel green synthesis and characterization of Ag NPs with its ultra-rapid catalytic reduction of methyl green dye, Appl. Surf. Sci. 290 (2014) 499-503. https://doi.org/10.1016/j.apsusc.2013.11.106

[56]　W. Qing, K. Chen, Y. Wang, X. Liu, M. Lu, Green synthesis of silver nanoparticles by waste tea extract and degradation of organic dye in the absence and presence of H_2O_2, Appl. Surf. Sci. 423 (2017) 1019-1024. https://doi.org/10.1016/j.apsusc.2017.07.007

[57]　S. Francis, S. Joseph, E.P. Koshy, B. Mathew, Microwave assisted green synthesis of silver nanoparticles using leaf extract of elephantopusscaber and its environmental and biological applications, Artif. Cells Nanomed. Biotechnol. 46 (2018) 795-804. https://doi.org/10.1080/21691401.2017.1345921

[58]　G. Arya, R.M. Kumari, N. Gupta, A. Kumar, R. Chandra, S. Nimesh, Green synthesis of silver nanoparticles using Prosopisjuliflora bark extract: reaction optimization, antimicrobial and catalytic activities, Artif. Cells Nanomed. Biotechnol. 46 (2018) 985-993. https://doi.org/10.1080/21691401.2017.1354302

[59]　T. N. J. I. Edison, Y.R. Lee, M.G. Sethuraman, Green synthesis of silver nanoparticles using Terminalia cuneata and its catalytic action in reduction of direct

yellow-12 dye, Spectrochim. Acta Part A 161 (2016) 122-129.
https://doi.org/10.1016/j.saa.2016.02.044

[60]   K. B. A. Ahmed, S. Subramaniam, G. Veerappan, N. Hari, A. Sivasubramanian, A. Veerappan, β-Sitosterol-d-glucopyranoside isolated from Desmostachyabipinnata mediates photoinduced rapid green synthesis of silver nanoparticles, *RSC. Adv.* 4 (2014) 59130-59136. https://doi.org/10.1039/C4RA10626A

[61]   N. K. R. Bogireddy, H. A. K.Kumar, B.K. Mandal, Biofabricated silver nanoparticles as green catalyst in the degradation of different textile dyes, J. Environ. Chem. Eng. 4 (2016) 56-64. https://doi.org/10.1016/j.jece.2015.11.004

[62]   C. T. N. J. I. Edison, R. Atchudan, M.G. Sethuraman, Y.R. Lee, Reductive-degradation of carcinogenic azo dyes using Anacardiumoccidentaletesta derived silver nanoparticles, J. Photochem. Photobiol. B 162 (2016) 604-610. https://doi.org/10.1016/j.jphotobiol.2016.07.040

[63]   Saravanan, R. Rajesh, T. Kaviarasan, K. Muthukumar, D. Kavitake, P.H. Shetty, Synthesis of silver nanoparticles using bacterial exopolysaccharide and its application for degradation of azo-dyes, Biotechnol. Rep. 15 (2017) 33-40. https://doi.org/10.1016/j.btre.2017.02.006

[64]   K. B. A. Ahmed, R. Senthilnathan, S. Megarajan, V. Anbazhagan, Sunlight mediated synthesis of silver nanoparticles using redox phytoprotein and their application in catalysis and colorimetric mercury sensing, J. Photochem. Photobiol. B 151 (2015) 39-45. https://doi.org/10.1016/j.jphotobiol.2015.07.003

[65]   K. Anand, K. Kaviyarasu, S. Muniyasamy, S.M. Roopan, R.M. Genga, A.A. Chuturgoon, Bio-synthesis of silver nanoparticles using agroforestry residue and their catalytic degradation for sustainable waste management, J. Clust. Sci. 28 (2017) 2279-2291. https://doi.org/10.1007/s10876-017-1212-2

[66]   K. Muthu, S. Priya, Green synthesis, characterization and catalytic activity of silver nanoparticles using Cassia auriculata flower extract separated fraction, Spectrochim. Acta Part A 179 (2017) 66-72. https://doi.org/10.1016/j.saa.2017.02.024

[67]   G. Manjari, S. Saran, T. Arun, S.P. Devipriya, A.V.B. Rao, Facile Aglaia clacagnoidea mediated synthesis of silver and gold nanoparticles: antioxidant and catalysis properties, J. Clust. Sci. 28 (2017) 2041-2056. https://doi.org/10.1007/s10876-017-1199-8

[68]   R. Vijayan, S. Joseph, B. Mathew, Green synthesis of silver nanoparticles using Nervaliazeylanica leaf extract and evaluation of their antioxidant, catalytic, and antimicrobial potentials, Particul. Sci. Technol. 37 (2019) 809-819. https://doi.org/10.1080/02726351.2018.1450312

[69]   S.A. Aromal, D. Philip, Green synthesis of gold nanoparticles using Trigonellafoenum-graecum and its size-dependent catalytic activity, Spectrochim. Acta A 97 (2012) 1-5. https://doi.org/10.1016/j.saa.2012.05.083

[70]   P. Dauthal, M. Mukhopadhyay, Prunusdomestica fruit extract-mediated synthesis of gold nanoparticles and its catalytic activity for 4-nitrophenol reduction,Ind. Eng. Chem. Res. 51 (2012) 13014-13020. https://doi.org/10.1021/ie300369g

[71]   S. Maity, I.K. Sen, S.S. Islam, Green synthesis of gold nanoparticles using gum polysaccharide of Cochlospermumreligiosum (katira gum) and study of catalytic activity,Physica E Low Dimens. Syst. 45 (2012) 130-134. https://doi.org/10.1016/j.physe.2012.07.020

[72]   IK Sen, K. Maity, S.S. Islam, Green synthesis of gold nanoparticles using a glucan of an edible mushroom and study of catalytic activity, Carbohydr. Polym. 91 (2013) 518-528. https://doi.org/10.1016/j.carbpol.2012.08.058

[73]   R. Majumdar, B.G. Bag, N. Maity, Acacia nilotica (Babool) leaf extract mediated size-controlled rapid synthesis of gold nanoparticles and study of its catalytic activity, Int. Nano Lett. 3 (2013) 53. https://doi.org/10.1186/2228-5326-3-53

[74]   C. Tamuly, M. Hazarika, M. Bordoloi, Biosynthesis of Au nanoparticles by Gymnocladusassamicus and its catalytic activity, Mater. Lett. 108 (2013) 276-279. https://doi.org/10.1016/j.matlet.2013.07.020

[75]   SS Dash, BG Bag, Synthesis of gold nanoparticles using renewable Punicagranatum juice and study of its catalytic activity, Appl. Nanosci. 4 (2014) 55-59. https://doi.org/10.1007/s13204-012-0179-4

[76]   K, Paul, B.G. Bag, K. Samanta, Green coconut (Cocos nucifera Linn) shell extract mediated size controlled green synthesis of polyshaped gold nanoparticles and its application in catalysis, Appl. Nanosci. 4 (2014) 769-775. https://doi.org/10.1007/s13204-013-0261-6

[77]   SS, Dash, R. Majumdar, A.K. Sikder, B.G. Bag, BKPatra, Saracaindica bark extract mediated green synthesis of polyshaped gold nanoparticles and its application in catalytic reduction, Appl. Nanosci.4 (2014) 485-490. https://doi.org/10.1007/s13204-013-0223-z

[78]   Y. Choi, M.J. Choi, S.H. Cha, Y.S. Kim, S. Cho, Y. Park, Catechin-capped gold nanoparticles: green synthesis, characterization, and catalytic activity toward 4-nitrophenol reduction, Nanoscale Res. Lett. 9 (2014) 103. https://doi.org/10.1186/1556-276X-9-103

[79]   A. Rajan, M. MeenaKumari, D. Philip, Shape tailored green synthesis and catalytic properties of gold nanocrystals, Spectrochim. Acta A 118 (2014) 793-799. https://doi.org/10.1016/j.saa.2013.09.086

[80]   Z. Gao, R. Su, R. Huang, W. Qi, Z. He, Glucomannan-mediated facile synthesis of gold nanoparticles for catalytic reduction of 4-nitrophenol, Nanoscale Res. Lett. 9 (2014) 404. https://doi.org/10.1186/1556-276X-9-404

[81]   SS Dash, BG Bag, P. Hota, Lantana camara Linn leaf extract mediated green synthesis of gold nanoparticles and study of its catalytic activity, Appl. Nanosci. 5 (2015) 343-350. https://doi.org/10.1007/s13204-014-0323-4

[82]   K. Anand, R.M. Gengan, A. Phulukdaree, A. Chuturgoon, Agroforestry waste Moringaoleifera petals mediated green synthesis of gold nanoparticles and their anti-cancer and catalytic activity, J. Ind. Eng. Chem. 21 (2015) 1105-1111. https://doi.org/10.1016/j.jiec.2014.05.021

[83]   N.K.R. Bogireddy, K.K.H. Anand, B.K., Mandal, Gold nanoparticles synthesis by Sterculiaacuminata extract and its catalytic efficiency in alleviating different organic dyes, J. Mol. Liq. 211 (2015) 868-875. https://doi.org/10.1016/j.molliq.2015.07.027

[84]   S. Wu, S. Yan, W. Qi, R. Huang, J. Cui, R. Su, Z He, Green synthesis of gold nanoparticles using aspartame and their catalytic activity for p-nitrophenol reduction, Nanoscale Res. Lett. 10 (2015) 213. https://doi.org/10.1186/s11671-015-0910-7

[85]   S. Das, B.G. Bag, R. Basu, Abromaaugusta Linn bark extract-mediated green synthesis of gold nanoparticles and its application in catalytic reduction, Appl. Nanosci. 5 (2015) 867-873. https://doi.org/10.1007/s13204-014-0384-4

[86]   J. Yu, D. Xu, H.N. Guan, C. Wang, L.K. Huang, Facile one-step green synthesis of gold nanoparticles using Citrus maxima aqueous extracts and its catalytic activity, Mater. Lett. 166 (2016) 110-112. https://doi.org/10.1016/j.matlet.2015.12.031

[87]   R. Majumdar, B.G. Bag, P. Ghosh, Mimusopselengi bark extract mediated green synthesis of gold nanoparticles and study of its catalytic activity, Appl. Nanosci. 6 (2016) 521-528. https://doi.org/10.1007/s13204-015-0454-2

[88]   M. Ramakrishna, D.R. Babu, RM. Gengan, S. Chandra, G.N. Rao, Green synthesis of gold nanoparticles using marine algae and evaluation of their catalytic activity, J. Nanostructure Chem. 6 (2016) 1-13. https://doi.org/10.1007/s40097-015-0173-y

[89]   K. Banerjee, V.R. Rai, Study on green synthesis of gold nanoparticles and their potential applications as catalysts, J. Clust. Sci. 27 (2016) 1307-1315. https://doi.org/10.1007/s10876-016-1001-3

[90]   J. Park, S.H. Cha, S. Cho, Y. Park, Green synthesis of gold and silver nanoparticles using gallic acid: catalytic activity and conversion yield toward the 4-nitrophenol reduction reaction, J. Nanopart. Res. 18 (2016) 166. https://doi.org/10.1007/s11051-016-3466-2

[91]   X. Zhang, Y. Qu, W. Shen, J. Wang, H. Li, Z. Zhang, S. Li, J. Zhou, Biogenic synthesis of gold nanoparticles by yeast Magnusiomycesingens LH-F1 for catalytic reduction of nitrophenols, Colloids Surf. A Physicochem. Eng. Aspects 497 (2016) 280-285. https://doi.org/10.1016/j.colsurfa.2016.02.033

[92]   A.G. Assefa, AAMesfin, M.L. Akele, A.K. Alemu, B.R. Gangapuram, V.Guttena, M. Alle, Microwave-assisted green synthesis of gold nanoparticles using Olibanum gum (Boswelliaserrate) and its catalytic reduction of 4-nitrophenol and hexacyanoferrate (III) by sodium borohydride, J. Clust. Sci. 28 (2017) 917-935. https://doi.org/10.1007/s10876-016-1078-8

[93]   S. Sunkari, B.R. Gangapuram, R. Dadigala, R. Bandi, M. Alle, V Guttena, Microwave-irradiated green synthesis of gold nanoparticles for catalytic and anti-bacterial activity, J. Anal. Sci. Technol. 8 (2017) 13. https://doi.org/10.1186/s40543-017-0121-1

[94]   N.K.R. Bogireddy, U. Pal, L.M. Gomez, V. Agarwal, Size controlled green synthesis of gold nanoparticles using Coffeaarabica seed extract and their catalytic performance in 4-nitrophenol reduction, RSC.Advan. 8 (2018) 24819-24826. https://doi.org/10.1039/C8RA04332A

[95]   C. Umamaheswari, A. Lakshmanan, NS. Nagarajan, Green synthesis, characterization and catalytic degradation studies of gold nanoparticles against congo red and methyl orange, J Photochem. Photobiol. B 178 (2018) 33-39. https://doi.org/10.1016/j.jphotobiol.2017.10.017

[96]   B.R. Gangapuram, R.Bandi, M. Alle, R. Dadigala, G.M. Kotu, V. Guttena, Microwave assisted rapid green synthesis of gold nanoparticles using Annona squamosa L peel extract for the efficient catalytic reduction of organic pollutants, J. Mol. Struct. 1167 (2018) 305-315. https://doi.org/10.1016/j.molstruc.2018.05.004

[97]   Z. Issaabadi, M. Nasrollahzadeh, S.M. Sajadi, Green synthesis of the copper nanoparticles supported on bentonite and investigation of its catalytic activity, J. Clean. Prod. 142 (2017) 3584-3591. https://doi.org/10.1016/j.jclepro.2016.10.109

[98]   S. Saran, G. Manjari, S.P. Devipriya, Synergistic eminently active catalytic and recyclable Ag, Cu and Ag-Cu alloy nanoparticles supported on TiO_2 for sustainable and cleaner environmental applications: A phytogenic mediated synthesis, J. Clean. Prod. 177 (2018) 134-143. https://doi.org/10.1016/j.jclepro.2017.12.181

[99]   M. Ismail, S. Gul, M.I. Khan, M.A. Khan, A.M. Asiri, S.B. Khan, Green synthesis of zerovalent copper nanoparticles for efficient reduction of toxic azo dyes congo red and methyl orange, Green Process. Synth. 8 (2019) 135-143. https://doi.org/10.1515/gps-2018-0038

[100] H.S. Devi, T.D. Singh, Synthesis of copper oxide nanoparticles by a novel method and its application in the degradation of methyl orange, Adv. Electron Electr. Eng. 4 (2014) 83-88.

[101] VNKalpana, P. Chakraborthy, V. Palanichamy, V.D Rajeswari, Synthesis and characterization of copper nanoparticles using Tridaxprocumbens and its application in degradation of bismarck brown, Analysis 10 (2016) 17.

[102] G. Manjari, S. Saran, T. Arun, A.V.B. Rao, S.P. Devipriya, Catalytic and recyclability properties of phytogenic copper oxide nanoparticles derived from Aglaia elaeagnoidea flower extract, J. Saudi Chem. Soc. 21 (2017) 610-618. https://doi.org/10.1016/j.jscs.2017.02.004

[103] B. Khodadadi, M. Bordbar, M. Nasrollahzadeh, Achilleamillefolium L. extract mediated green synthesis of waste peach kernel shell supported silver nanoparticles: Application of the nanoparticles for catalytic reduction of a variety of dyes in water, J. Colloid Interface Sci.493 (2017) 85-93. https://doi.org/10.1016/j.jcis.2017.01.012

[104] M. Qasem, R. El Kurdi, D.Patra, Green Synthesis of Curcumin Conjugated CuO Nanoparticles for Catalytic Reduction of Methylene Blue, ChemistrySelect 5 (2020) 1694-1704. https://doi.org/10.1002/slct.201904135

[105] A.J. Kora, L. Rastogi, Green synthesis of palladium nanoparticles using gum ghatti (Anogeissuslatifolia) and its application as an antioxidant and catalyst, Arab. J. Chem. 11 (2018) 1097-1106. https://doi.org/10.1016/j.arabjc.2015.06.024

[106] A.J. Kora, L. Rastogi, Catalytic degradation of anthropogenic dye pollutants using palladium nanoparticles synthesized by gum olibanum, a glucuronoarabinogalactan

Materials Research Forum LLC
https://doi.org/10.21741/9781644901830-1

biopolymer, Ind. Crops Prod. 81 (2016) 1-10.
https://doi.org/10.1016/j.indcrop.2015.11.055

[107] G. Li, Y. Li, Z. Wang, H. Liu, Green synthesis of palladium nanoparticles with carboxymethyl cellulose for degradation of azo-dyes, Mater. Chem. Phys. 187(2017) 133-140. https://doi.org/10.1016/j.matchemphys.2016.11.057

[108] M. Bordbar, N. Mortazavimanesh, Green synthesis of Pd/walnut shell nanocomposite using Equisetum arvense L. leaf extract and its application for the reduction of 4-nitrophenol and organic dyes in a very short time, Environ. Sci. Pollut. Res. 24 (2017) 4093-4104. https://doi.org/10.1007/s11356-016-8183-y

[109] C. Garai, S.N. Hasan, A.C. Barai, S. Ghorai, S.K. Panja, B.G. Bag, Green synthesis of Terminalia arjuna-conjugated palladium nanoparticles (TA-PdNPs) and its catalytic applications, J. Nanostructure Chem, 8(2018) 465-472. https://doi.org/10.1007/s40097-018-0288-z

[110] M. Nasrollahzadeh, S.M. Sajadi, M. Maham, Green synthesis of palladium nanoparticles using Hippophaerhamnoides Linn leaf extract and their catalytic activity for the Suzuki–Miyaura coupling in water, J. Mol. Catal. A: Chem. 396 (2015) 297-303. https://doi.org/10.1016/j.molcata.2014.10.019

[111] M. Nasrollahzadeh, S.M. Sajadi, A. Rostami-Vartooni, M. Alizadeh, M. Bagherzadeh, Green synthesis of the Pd nanoparticles supported on reduced graphene oxide using barberry fruit extract and its application as a recyclable and heterogeneous catalyst for the reduction of nitroarenes, J. Colloid Interface Sci. 466 (2016) 360-368. https://doi.org/10.1016/j.jcis.2015.12.036

[112] M. Khan, M. Khan, M. Kuniyil, S.F. Adil, A. Al-Warthan, H.Z. Alkhathlan, M. R. H. Siddiqui, Biogenic synthesis of palladium nanoparticles using Pulicariaglutinosa extract and their catalytic activity towards the Suzuki coupling reaction, *Dalton Trans.* 43 (2014) 9026-9031. https://doi.org/10.1039/C3DT53554A

[113] M. Nasrollahzadeh, S.M. Sajadi, Pd nanoparticles synthesized in situ with the use of Euphorbia granulate leaf extract: Catalytic properties of the resulting particles, J. Colloid Interface Sci. 462 (2016) 243-251. https://doi.org/10.1016/j.jcis.2015.09.065

[114] M. Khan, G.H. Albalawi, M.R. Shaik, M. Khan, S.F. Adil, M. Kuniyil, M. R. H. Siddiqui, Miswak mediated green synthesized palladium nanoparticles as effective catalysts for the Suzuki coupling reactions in aqueous media, Saudi Chem. Soc. 21(2017) 450-457. https://doi.org/10.1016/j.jscs.2016.03.008

[115] R. Venu, T.S. Ramulu, S. Anandakumar, V.S. Rani, C.G. Kim, Bio-directed synthesis of platinum nanoparticles using aqueous honey solutions and their catalytic applications, Colloids Surf. A Physicochem. Eng. Aspects384 (2011) 733-738. https://doi.org/10.1016/j.colsurfa.2011.05.045

[116] D.S. Sheny, D. Philip, J. Mathew, Synthesis of platinum nanoparticles using dried Anacardiumoccidentale leaf and its catalytic and thermal applications, Spectrochim. Acta A 114 (2013) 267-271. https://doi.org/10.1016/j.saa.2013.05.028

[117] S. Pandey, S.B. Mishra, Catalytic reduction of p-nitrophenol by using platinum nanoparticles stabilized by guar gum, Carbohydr. Polym. 113 (2014) 525-531. https://doi.org/10.1016/j.carbpol.2014.07.047

[118] P. Dauthal, M. Mukhopadhyay, Biofabrication, characterization, and possible bio-reduction mechanism of platinum nanoparticles mediated by agro-industrial waste and their catalytic activity, J. Ind. Eng. Chem. 22 (2015) 185-191. https://doi.org/10.1016/j.jiec.2014.07.009

[119] W. Ye, J. Yu, Y Zhou, D. Gao, D. Wang, C. Wang, D. Xue, Green synthesis of Pt–Au dendrimer-like nanoparticles supported on polydopamine-functionalized graphene and their high performance toward 4-nitrophenol reduction, Appl. Catal. B Environ. 181 (2016) 371-378. https://doi.org/10.1016/j.apcatb.2015.08.013

[120] VSRamkumar, A. Pugazhendhi, S. Prakash, N.K. Ahila, G. Vinoj, S. Selvam, G. Kumar, E. Kannapiran, R.B. Rajendran, Synthesis of platinum nanoparticles using seaweed Padinagymnospora and their catalytic activity as PVP/PtNPs nanocomposite towards biological applications, Biomed. Pharmacother. 92 (2017) 479-490. https://doi.org/10.1016/j.biopha.2017.05.076

[121] R. Dobrucka, Biofabrication of platinum nanoparticles using Fumariaeherba extract and their catalytic properties, Saudi J. Biol. Sci. 26(2019) 31-37. https://doi.org/10.1016/j.sjbs.2016.11.012

[122] G.E. Hoag, J.B. Collins, J.L. Holcomb, J.R. Hoag, M.N. Nadagouda, R.S. Varma, Degradation of bromothymol blue by 'greener' nanoscale zero-valent iron synthesized using tea polyphenols, J. Mater. Chem. 19 (2009) 8671–8677. https://doi.org/10.1039/b909148c

[123] E.C. Njagi, H. Huang, L. Stafford, H. Genuino, H.M. Galindo, J.B. Collins, G.E. Hoag, S.L. Suib, Biosynthesis of iron and silver nanoparticles at room temperature using aqueous Sorghum bran extracts, Langmuir 27 (2011) 264–271. https://doi.org/10.1021/la103190n

[124] T. Shahwan, S. Abu Sirriah, M. Nairat, E. Boyacı, A.E. Ero ˇglu, T.B. Scott, K.R. Hallam Green synthesis of iron nanoparticles and their application as a fenton-like catalyst for the degradation of aqueous cationic and anionic dyes, Chem. Eng. J. 172 (2011) 258–266. https://doi.org/10.1016/j.cej.2011.05.103

[125] L. Huang, X. Weng, Z. Chen, M. Megharaj, R. Naidu, Synthesis of iron-based nanoparticles using Oolong tea extract for the degradation of malachite green, Spectrochim. Acta A 117 (2013) 801–804. https://doi.org/10.1016/j.saa.2013.09.054

[126] Y. Kuang, Q. Wang, Z. Chen, M. Megharaj, R. Naidu, Heterogeneous fenton-like oxidation of monochlorobenzene using green synthesis of iron nanoparticles, J. Colloid Interface Sci. 410 (2013) 67–73. https://doi.org/10.1016/j.jcis.2013.08.020

[127] L. Huang, F. Luo, Z. Chen, M. Megharaj, R. Naidu, Green synthesized conditions impacting on the reactivity of Fe NPs for the degradation of malachite green, Spectrochim. Acta A 137 (2015) 154–159. https://doi.org/10.1016/j.saa.2014.08.116

[128] F. Luo, Z. Chen, M. Megharaj, R. Naidu, Biomolecules in grape leaf extract involved in one-step synthesis of iron-based nanoparticles, RSC. Adv. 4 (2014) 53467–53474. https://doi.org/10.1039/C4RA08808E

[129] Truskewycz, R. Shukla, AS Ball, Iron nanoparticles synthesized using green tea extracts for the fenton-like degradation of concentrated dye mixtures at elevated temperatures, J. Environ. Chem. Eng. 4 (2016) 4409-4417. https://doi.org/10.1016/j.jece.2016.10.008

[130] M. Khoshnamvand, C. Huo, J. Liu, Silver nanoparticles synthesized using Allium ampeloprasum L. leaf extract: characterization and performance in catalytic reduction of 4-nitrophenol and antioxidant activity. J. Mol. Struct. 1175 (2019) 90-96. https://doi.org/10.1016/j.molstruc.2018.07.089

Bioinspired Nanomaterials for Energy and Environmental Applications   Materials Research Forum LLC
Materials Research Foundations **121** (2022) 39-82   https://doi.org/10.21741/9781644901830-2

Chapter 2

Bio-Inspired Metal Oxide Nanostructures for Photocatalytic Disinfection

Muthuraj Arunpandian[1], Tammineni Venkata Surendra[2], Norazriena Yusoff[3],
Saravana Vadivu Arunachalam[1,4,*]

[1]Department of Chemistry, International Research Centre, Kalasalingam Academy of Research and Education, Krishnankoil - 626 126, Virudhunagar, Tamil Nadu, India

[2]Department of Chemistry, Chaitanya Bharathi Institute of Technology (A), Gandipet, Hyderabad, Telangana 500075, India

[3]Photonics Research Centre, University of Malaya, Kuala Lumpur – 50603, Malaysia

[4]Department of chemistry, Saveetha School of Engineering, Saveetha Institute of Medical and Technical Sciences, Chennai, Tamil Nadu – 602105, India

* drarunachalam.s@gmail.com

Abstract

Interest in photocatalytic disinfection synthesis has increased in recent years with the use of different semiconductor photoreceptors. While much attention has been given to the photocatalytic inactivation process, researchers have shifted to focusing on bio-inspired metal oxide materials for photocatalytic inactivation in recent years. Bio-inspired metal oxide photocatalysts have unique advantages with special emphasis being placed on its highly earth abundance, economic cost of production, eco-friendliness, simple structure and easy to synthesize. Besides that, bio-inspired metal oxide photocatalysts has also been applied extensively for the development of emerging areas, such as environmental as well as energy materials. Today, the development of simple and inexpensive bacterial disinfection technology to addresses the peril of waterborne disease in the emerging areas has grown rapidly. This chapter proposes an analysis of recent research activities that involved the use of bio-inspired photocatalytst for the disinfection of water under light radiation. Various nano-structured photocatalytic materials like titanium dioxide (TiO_2), zinc oxide (ZnO), iron oxide (Fe_2O_3), nickel oxide (NiO), etc., are introduced. Material and various bacterial pathogens, photocatalytic and pathogens disinfection mechanism are described in detail. Finally, the progress of novel bio-inspired photocatalysts for the disinfection applications is discussed at the end of this chapter.

Bioinspired Nanomaterials for Energy and Environmental Applications   Materials Research Forum LLC
Materials Research Foundations **121** (2022) 39-82   https://doi.org/10.21741/9781644901830-2

Keywords

Bio-Inactivation, Metal Oxides, Bio-Inspired Materials, Light Irradiation, Microorganisms, Green Synthesis, Disinfection

Contents

Bio-Inspired Metal Oxide Nanostructures for Photocatalytic Disinfection.39

1.   **Introduction**..**41**

2.   **Photocatalytic method of disinfection**..**44**

3.   **Bio-inspired photocatalysts**...**44**

    3.1   Bio-inspired nanostructures photocatalysts..45

    3.1.1 Biopolymers based photocatalysts...45

    3.1.2 Biochars based photocatalysts..45

    3.1.3 Immobilized enzymes, peptides and other biomolecules
           based photocatalysts 46

    3.1.4. Bio-inspired photocatalyst via green synthesis 47

4.   **Bioinspired metal oxide nanostructures for
photocatalytic disinfection**...**48**

    4.1   Photocatalytic disinfection of biological pathogens using
           g-C_3N_4-based photocatalysts 48

    4.2   Photocatalytic disinfection using NiO nano-rods.............................48

    4.3   Photocatalytic inactivation of bacteria's using FeONPs
           alone and FeONPs incorporated cotton fabrics materials 51

    4.4   Photocatalytic inactivation using nano-flower ZnO catalysts..........52

    4.5 TiO_2 nanoparticles to inactivate pathogens in water 53

    4.6   Photocatalytic inactivation of pathogens in water using
           extracellular biosynthesis of cobalt ferrite nanoparticles 54

    4.7   Photocatalytic inactivation using different doses of g-C_3N_4
           photocatalysts 55

    4.8   Photocatalytic disinfection using C. abyssinica tuber extract
           mediated synthesized ZnO nanoparticles 56

    4.9   Photocatalytic inactivation analysis using Biosynthesis of
           Ag deposited phosphorus and sulfur co-doped g-C_3N_4 (Ag-
           PSCN) nanocomposite..58

4.10   Photocatalytic disinfection of pathogens in water using
        Nd_2WO_6/ZnO incorporated on GO (NWZG) nanocomposite 60

4.11   Visible-light-driven photocatalytic disinfection of bacteria's
        using Zea mays L. dry husk mediated bio-synthesized
        copper oxide nanoparticles 60

4.12   Efficient photocatalytic disinfection of Escherichia coli
        O157:H7 using C70-TiO_2 hybrid material under visible
        light irradiation 61

4.12.1 Photocatalytic disinfection mechanism 61

4.13   Antibacterial assay on SnO_2 doped GO and CNT under
        visible light irradiation..62

4.14   Photocatalytic inactivation on ZnO nanoparticle under
        visible light irradiation..63

4.15   Photocatalytic disinfection of bacterial pathogens using
        MgO nanostructures..65

Conclusions..**68**

References **68**

1.   Introduction

Recent advances in nanotechnology have verified the major roles played by nanomaterials. Generally, nanotechnology refers to the preparation, spectral analysis, and material uses that are within the nanometer (nm) scale range (1 to 100 nm). The material possess at least one external dimension measured in this range is known as nanomaterial. Nanomaterial can be broadly divided into two types: These include (i) C- based nanoparticles (NPs) like fullerenes, (ii) inorganic particles include noble metals (gold (Au), silver (Ag) and platinum (Pt)), magnetic NPs (iron (Fe), cobalt (Co) and nickel (Ni)), and semiconductors (oxides of titanium (Ti), zinc(Zn) and cadmium (Cd)) to mention a few. The biosynthesis of NPs, in general, involves this approach in which biomolecules mediating reduction processes and stabilizing nanoparticles. In the previous literature survey on the preparation of NPs, the chemical or physical techniques are typically being used. However, these techniques having some issues involving the use of high temperatures, radiation, and hazardous materials in the preparation process, as well as the need for–specialized equipment and consumed energy. Due to these drawbacks, bio systems have been used as efficient substitute to the rapid, cost-effectual and "Eco-friendly" synthesis of NPs. Generally, bio systems including plants and microscopic organisms such as viruses and bacteria consist of variation of biomolecules with reduced properties [1-10]. With that said, the bio system

Bioinspired Nanomaterials for Energy and Environmental Applications   Materials Research Forum LLC
Materials Research Foundations **121** (2022) 39-82   https://doi.org/10.21741/9781644901830-2

has the natural ability to reduce metallic ions into neutral atoms in non-hazardous and toxic ways.

In recent years, the world has faced serious environmental problems in terms of environmental contamination, changing in climate and infections [11-13]. Some of bacteria, cyanobacteria, viruses, prions, protozoa, cysts, fungi, helminthes and rickettsia, etc., prove to be dangerous to humans and pose a serious threat to aquatic systems. The microbial profile of bacterial pathogens is given below:

Escherichia Coli

Escherichia coli, also termed as E. coli, belong to maximize and diverse group of microbial organisms under the type of bacteria. While almost all strains of E. coli are innocuous, others are pathogenic. Some types of E. coli can source diarrhea; otherwise extras are urinary and respiratory tract disease contagions such as pneumonia, and have the potential to cause other diseases. Some other types of E. coli are used as markers of water pollution because they are found in water, which is not harmful to them, but rather to indicate the degree of water pollution.

PseudomonasAeruginosa

Pseudomonas is a kind of bacteria generally determined in water and soil. Of the various kind of Pseudomonas, one of the exceedingly frequent sickness contagions in human beings is referred to as Pseudomonas aeruginos. It may also purpose disorder contagions inside the human blood, lung troubles together with pneumonia, or maybe other elements of the frame after surgical incidents. These microorganisms are resistant to many drugs. In such circumstances, the drug fails to serve the meant motive.

Staphylococcus Aureus

Staphylococcus aureus, a kind of germ that is present in about 30% of people's nose. Usually it does no harm, and sometimes it causing epidemic problems. These infections can be serious or dangerous, such as bacteria or sepsis when bacteria spread into the bloodstream; Pneumonia, which often impacts human beings with lung problems; Endocarditis is a physiological condition that influences the coronary heart valves, that may lead to coronary heart failure or stroke; Osteomyelitis, a condition characterized by infection in the bones, is caused by staphylococcus bacteria in the bloodstream, or by pancreatic injury or intracranial irregularities in the foot.

Bacillus Anthracis

Bacillus anthracis is the largest and is gram-positive bacterium with 1 to 1.2 µm wide and 3 to 5µm in length. The bacteria can growth under two environments which are an oxygenated environment known as aerobic or in the absence of free oxygen called as anaerobic. It is the bacterium that causes anthrax. Anthrax differs from other bacteria in the sense that they are inactive, called "spores." Anthrax spores have been inactive in the soil for years, perhaps decades. The spores are located in animal carcasses and faeces, and animal merchandise consisting of hides and wool. Some animals are very resistant to anthrax. The spores can become active bacteria under appropriate conditions.

Salmonella Typhi

Salmonella typhi (S. typhi) is a bacterial strain that affect the intestines and blood. This disease is called typhoid fever. In many developing countries, S.Tifi and S. Paratyphi are common bacteria that can be found due to the very poor sewage and water treatment systems. Salmonella typhi and salmonella paratyphi, cause typhoid fever and paratyphoid fever, respectively, which are one of the life-threatening diseases.

Candida Albicans

Candida albicans is a yeast-like fungus that can be detected in humans gut to be specific. The number of these organisms is controlled by the human body and when there is imbalance, it can cause intestinal candidiasis or yeast infections in other parts of the body. Unplanned resources and the industrial revolt have brought about renewable power sources and the environment crisis [14]. These ubiquitous problems around the world have worsened in developing a part of the world as a result of the Industrial Revolution and poor surveillance and regulation [15]. Resource management, pollution reduction and disinfection technologies are well prepared for a variety of environmental issues [16, 17].

Further, the shortage of fossil fuels or renewable strength sources has heightened medical interest in bio-based materials and inexperienced technologies to reduce pollutants and green electricity generation. Excessive heavy metallic infection is of the subject as organisms with liver damage, kidney failure; breathing issues and heavy steel exposure have expanded [18]. Elimination of metals via absorption is continually the desired approach; however, it will increase the chance of secondary pollution [19]. This can be overcome without difficulty with the aid of reduction into non-toxic or low poisonous ions. Due to the excessive efficiency and complete mineralization ability, photocatalysis has become the pioneer in the superior treatment of water and disinfection [20-22]. However, numerous new production waste, from medical, plastics industry, agriculture, and housing have given rise to the sizeable poisonous pollutants. On an extensive line, they are classified as medicines, non-public care merchandise, and endocrine disorders [23, 24].

New age pollutants have challenged current "superstar" photocatalysis products and techniques. Consequently, extra superior photosynthesis and improvement in current photocatalysts technology are wished. Mild absorption in a normal photocatalytic mechanism results in the formation of electron-hollow pairs, that generate hydroxyl and superoxide ion radicals, thus induced the degradation of organic pollutants [25]. It is believed that the electrons reduce pollution as heavy metallic ions and mineral ions. For instance, the photovoltaic electrons bring heavy metal toxicity in Cr (VI) to non-poisonous Cr (III) through photoreduction [26]. –Whilst photosynthesis is supported in substances with excessive region, activity, and porosity, they perform better with higher dispersion and contamination [27].

2.   Photocatalytic method of disinfection

The process of killing pathogenic microorganisms is called disinfection. There are many procedures for disinfection, including chlorination, ozonation, ultraviolet light and photocatalysis. Photo-catalytically non-materials act as catalyst for degrading organic compounds, contaminating them, and disinfecting pathogens. These materials assist in the creation of reactive oxygen species (ROS) in the disinfection process, effectively carrying out the intended action. Their performance arises from the narrow band gap energy for hole-electron recombination. Their reusability and effectiveness make them an attractive choice for selection.

3.   Bio-inspired photocatalysts

Reliable toxicity of nanomaterials, the new realm of nanomaterials is called bio-inspired materials. The bio-inspired materials can be categorized into two: biomimics and bio templates. Generally, this type of material used the biological methods to synthesized nanomaterials and is environmentally friendly and highly adaptive. Bio-inspired nanomaterials establish low toxicity, high selectivity and specific target and disinfectant properties [28]. Fig. 1 illustrated the different types of bio-inspired photocatalysts.

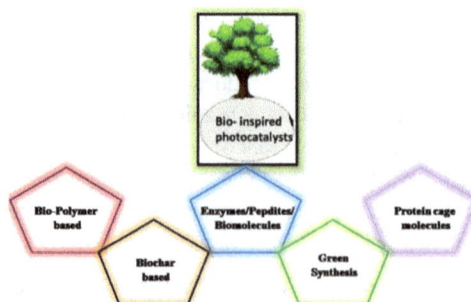

Fig. 1. Types of bio-inspired photocatalysts

In the last few years, major work has been devoted to the development of innovative materials by integrating the photoinduced properties of metallic oxide nanoparticles and the special surface moisture content of biomaterials. This research path has drawn vast medical attention to primary research and practical applications inside the fields of power, surroundings, medicine, industry, and different fields. The following sections of this review will focus on the latest developments on the mechanism, fabrication, and application of bio-inspired nanomaterials with special moisture content.

3.1   Bio-inspired nanostructures photocatalysts

3.1.1   Biopolymers based photocatalysts

Manufactured polymers are significant as a component of photosynthesis and the catalyst for a reasonable turn of events and ecological consideration is the thing that drives us to "come back to nature". Bio-polymers, for example, cellulose, chitosan, alginates, gelatin, cyclo-dextrins, normal gums, agar, gelatin, and collagen have pulled in materials for some ecological and organic applications in light of their properties and bounty. Adaptable properties like profoundly proficient execution, enormous surface, poisonousness, biological activity, biocompatibility, adaptability, and the simplicity of preparing have make it valuable for the used in various applications including in food bundling [29], clinical industry [30], biofuels [31], water treatment [32], vitality stockpiling [33]] and bio-medicinal applications [34].

Bio-polymers based phenomenal nanocomposite substances like TiO_2/oxidized CS-GLA [35] alginate/carbon nanotube/carbon speck/fluoroapatite/TiO_2 [36], $ZnFe_2O_4$@CMC [37], Bi_2WO_6–TiO_2/starch [38], chitosan-La^{3+}-graphite [39], TiO_2/calcium alginate [40], ZnO/Compact discs QDs installed crosslinked chitosan [41], etc. Among the various types of bio-polymer, Chitosan (CS) which is one of the family members of polysaccharides has received considerable attention due to its valuable properties such as low poisonousness; immoderate biodegradability and biocompatibility. Therefore, it has been extensively used for meals, biomedical, splendour factories, and environmental packaging [42].

3.1.2   Biochars based photocatalysts

Bio-based materials in which utilized in environmental applications are carbon-based products that has been extracted from bio-sources. Biomass has super potential in terms of adsorption, biofuel manufacturing, electricity production, and environmental clean-up [43, 44]. Biochar is an appealing and cheaper carbon-wealthy material obtained from biogas and synthetic by pyrolysis under excessive strain finite oxygen or hydrothermal carbonation [45-47]. In biology, biomass is typically composed of cellulose, hemicellulose, and lignin [48].

Biochar and its modified or functional forms have become ideal candidates to be applied in environmental toxicity, CO_2 capture, soil migration and enrichment, nitrogen determination, energy storage and support for various materials. Carbonaceous materials for instance biochar, hydrochar, charcoal are activated due to their versatility to eliminate various organic pollutants such as heavy metals, ammonia, nitrates and phosphates [51, 52]. In addition, biochar-based nanomaterials are regarded as useful catalysts for different acid catalytic reactions such as the evaluation of organic acids, the oscillation of alcohols, and the hydration of biomass [53].The subsequent traits of biochar are very appropriate for green photocatalysis method, for example:

    i.   Coupling nanomaterials, biocompatibility and non-toxicity

    ii.   Chemical and thermal balance

    iii.   Convertible porosity and features, large area over the surface of the material

    iv.   Shuttling or facilitation of electrons, Electron sink or reservoir

    v.   The adsorption potential value is very high

    vi.   Reduce the band energy value

Fu et al [57] prepared $MnFe_2O_4$/biochar composites i.e.$MnFe_2O_4$/MX (in which X is biochar; the biochar derived from corn stems, leaves and cores respectively) using a solvothermal direction. Wang *et al.* [58] have successfully synthesized nanocomposites between ZnO and biochar by employed the of two routes- reduction and precipitation strategies. Chen and co-workers [59] have prepared the ZnO doped biochar via the simultaneous pyrolysis of biomass and zinc oxide raw materials.

3.1.3 Immobilized enzymes, peptides and other biomolecules based photocatalysts

From numerous environmental and bio-catalytic programs, many attempts have been focused on the immobilization of enzymes, peptides, microbes, and numerous extraordinary organic molecules. Immobilization of enzymes has been started in 1916 [60] and it has been extensively used in several applications especially on diverse industries associated with fabric [61], meals [62], starch conversion [63], paper industries [64] and detergent industries [65]. Several merits had been said with reference to the use of these materials for waste treatment [66] in particular for wastewater purification [67] and solid waste treatment. The immobilization can be achieved using several method including physical adsorption, entrapment into polymer or gel matrix, and crosslinking through covalent bonds [68]. In brief, foremost optimization and method are crucial to maintain the balance of the enzyme in which the conditions should be changed in keeping with the desired natural capabilities [69]. It is worth noting that some of the pollutants obtained from fossil fuels cannot be discarded resulted from their excessive hydrophobicity and solubility in water, hence remains within our environment. Typically, these contaminants possess very slow rate of microbial biodegradation [70], therefore, the use of enzyme remedy is an attractive manner to supply better effects. For instance, the enzymes in particular the peroxidases and lactase have offer a unique way to solve the disposal of herbal pollutants issues [71]. Peroxidases were most frequently utilized in numerous commercial enterprise applications due to their merits of excellent bio-catalytic activity, easy availability, excessive balance, and broader substrate specificity [72]. Other than that, peroxidases are also capable of oxidizing diverse natural pollutants which consist of dyes, phenols, pesticides, and metals. In the work reported by Paola *et al.* [73], they have developed a photoactive machine using titania and enzyme soybean peroxidase (TiO_2-Soyabean peroxidase) by incorporating into the polymer matrix. In another report, Chen *et al.* [74] have synthesized imidazole–functionalized g-C_3N_4 and axially coordinated it with the main for photodegradation of phenols and antibiotics.

3.1.4. Bio-inspired photocatalyst via green synthesis

Fig. 2 shows the numerous greener methods that can be employed to synthesize nanomaterials. These techniques normally use greener materials as decreasing dealers, capping dealers, stabilizers and supports. Following are the routes which can be used for greener synthesis:

    i.   Plant extracts-consists of stems, leaves, flowers and roots

    ii.   Natural oils, biomolecules as biopolymers and herbal gums

    iii. Micro-organisms

Those strategies promise to keep away from secondary contaminants, ambient response conditions, bio-beneficial strategies, and coffee strength consumption. Using green composite nanomaterials is properly applicable to photosynthetic environmental toxicity.

Recently, a new approach is proposed to extracts the Fe_3O_4 nanoparticles from Kappaphycus alvarezii plant via the hot plate combustion technique. [75]. To date, many bio-inspired photocatalysts had been diagnosed can be prepared under the ultra violet or visible light such as TiO_2, ZnO, SnO_2, Fe_2O_3, $BiVO_4$, Cu_2O, $CdIn_2S_4$ and Ag_3PO_4, and many others.

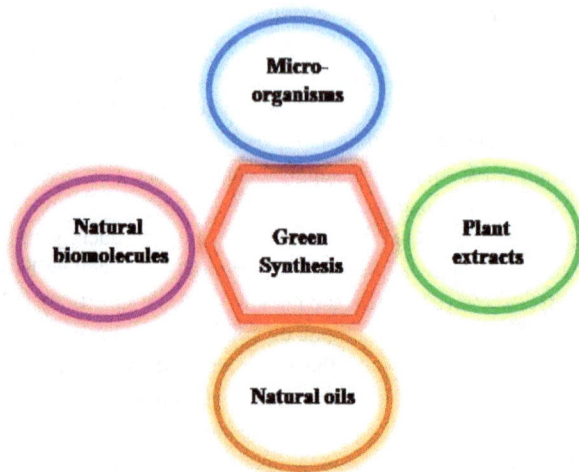

Fig. 2 Green synthetic routes to prepare the bio-inspired nanomaterials

4.   Bioinspired metal oxide nanostructures for photocatalytic disinfection

4.1   Photocatalytic disinfection of biological pathogens using g-C₃N₄-based photocatalysts

The photocatalytic bactericidal outcomes of g-C_3N_4 in opposition to E. coli was first discovered by Huang *et al.* [76], who have synthesized mesoporous g-C_3N_4 photocatalysts by means of the self-condensation of cyanamide with the assistance of a silica template. It is worth noting that the E. coli K-12 in the water which contain g-C_3N_4 have been effectively killed underneath the visible light irradiation. However, no disinfection happens in light without the presence of photocatalysts and darkish controls without mild irradiation. It became additionally discovered that the inactivation efficiency turned into appreciably influenced through the surface properties of g-C_3N_4. Li et al. have prepared high photocatalytic material g-C_3N_4/TiO_2 was effectively inactivate the bacterial pathogens using visible light driven and the photocatalyst material are synthesized by a facile hydrothermal- calcination route.

4.2   Photocatalytic disinfection using NiO nano-rods

It has been reported by Ezhilarasi and co-workers that the prepared nickel oxide nanorods (NiO NRs) display a top notch impact over gram positive, aside from gram negative bacterial strains [77]. Fig. 3 demonstrated the pattern of the inhibition zones that arise from the active functioning NiO NRs over microorganisms. It is an undeniable fact that NiO NRs are pretty powerful in opposition to gram positive pathogens compared to gram negative pathogens, resulted from the changes in susceptibility which rely upon the difference in the tolerance by oxidative stress. Besides that, other factors influencing the difference includes the composition of the cell, functionality, metabolism and the extent of contact between the organism and NiO NRs. Normally, gram positive bacteria carry one cytoplasmic membrane together with multi-peptidoglycan layers, while the gram negative bacteria consist of a complicated semi permeable membrane. Therefore, the damage within the semi permeable membrane of gram positive bacterial traces takes place effortlessly [77]. The completed antibacterial potential could be clarified with the aid of several factors such as size and shape of the particles, in addition to particular vicinity and purity of the material and particular surface area. The NiO NRs in small particle size will pass through the semi permeable membrane effortlessly and resulting in the damage of the cell due to intrude in the intracellular metabolism of calcium (II) ions.

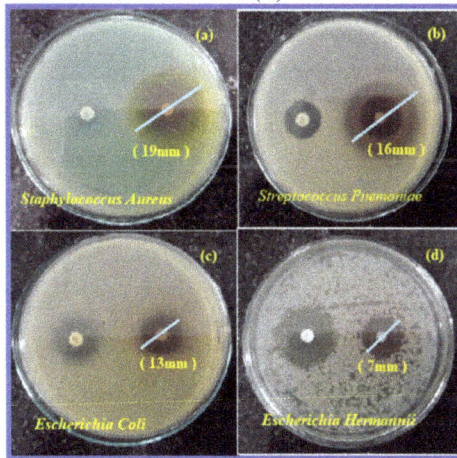

Fig. 3 Photographs displaying the inhibition zones of antibacterial activity induced by NiO nanorod: (a) Staphylococcus aureus - 19 mm (b) Streptococcus pnemoniae - 16 mm (c) Escherichia coli - 13 mm (d) Escherichia hermannii - 7 mm, (Reprinted with permission from Ref. [77] Copyright 2020, Elsevier).

Besides that, the large surface area owned by NiO NRs will further help to bind the bacterial additives. It has been reported that high interaction occurred on the particles with reduced size as a results of larger surface area, therefore increase its potential as antibacterial compared to larger particles [78]. It should be pointed out that NiO NRs have the high chances to bind with the functional groups provided by the proteins, hence inducing their denaturation and subsequently the death of cell [79]. Apart from that, the interaction of NiO NRs with sulphur and phosphorous consisting components such as DNA will cause the cell death, thus damage in bacterial cell [80].

The precise mechanism of antibacterial interest imparted by using metallic oxide nanomaterials have not been well described yet, however, few factors are ascribed to cause the activity. Among the factors, the production of reactive oxygen species (ROS) induced by nanomaterials has become the main factor that triggered the antibacterial activity [81, 82]. This is because it stimulates an excessive damage to lipids, proteins, amino acids, DNA and mitochondria [83]. Fig. 4 illustrated the schematic diagram of the antibacterial activity exhibited by NiO NRs due to the generation of ROS. The process of generating the ROS is started when the photoenergy possess by the metallic oxide nanoparticle is higher than its bandgap (Eg), thus leading to the formation of photoexcited electrons and holes. Next, the different ROS are formed as the molecule of oxygen has been absorbed by the electrons while the holes interacted with hydroxyl ions and water. In the effective antibacterial activity, the hydroxide radical plays an importants role as the dominant species while the Hydroxide radical ($\bullet$OH) and superoxide anion ($O^{2\bullet-}$), hydrogen peroxide (H_2O_2) are corresponds to various ROS. Besides that, other factors that have significantly affecting the antibacterial activity are the particle size of the material and the capability of metal oxides to release metal ions [84].

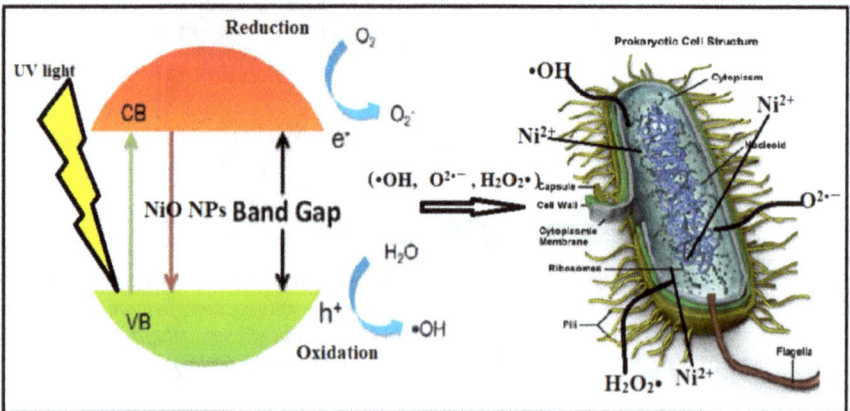

Fig. 4 The illustration of the formation of ROS via light irradiation on NiO NRs to restrain the bacterial activity. (Reprinted with permission from Ref. [77] Copyright 2020, Elsevier)

4.3   Photocatalytic inactivation of bacteria's using FeONPs alone and FeONPs incorporated cotton fabrics materials

The agar disc diffusion technique has been utilized in order to test the antibacterial activity of FeONPs alone and FeONPs incorporated fabrics on the Gram negative (E. coli, K. pneumoniae) and Gram positive bacteria (S. aureus). As can be seen in Fig. 5 (a-d) and Fig. 6 (a-c), the FeONPs alone and FeONPs incorporated cotton fabrics exhibits effective bactericidal activity. It has been verified that both materials demonstrated an efficient antibacterial activity upon tested on Gram negative bacteria particularly E. coli, whilst other pathogens such as K. pneumoniae and S. aureus are less sensitive to FeONPs as compared to E. coli. Interestingly, no zone of inhibition is appeared when the disc and cotton fabrics without FeONPs are tested towards all the three bacterial pathogens. It is believed that the incorporation of FeONPs on cotton fabrics may led to the disruption of the cell wall and further caused the cell death [85].

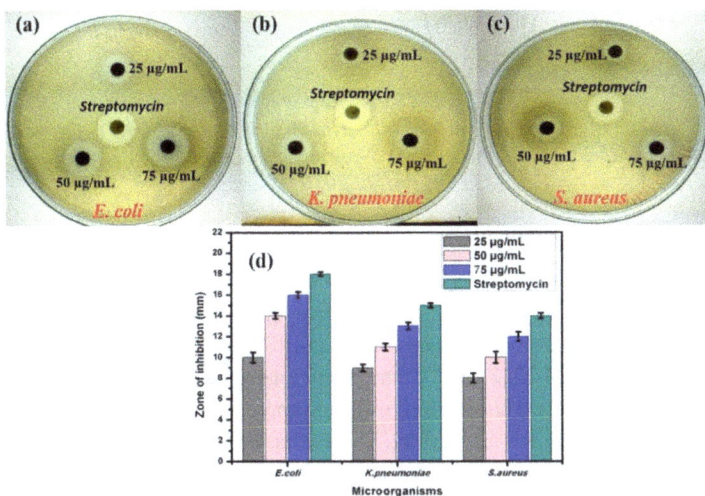

Fig. 5 The zone inhibition for various bacterial strains by FeONPs: (a) E. coli, (b) K. pneumoniae, (c) S. aureus and (d) zone of inhibition in mm (Reprinted with permission from Ref. [85] Copyright 2020, Elsevier)

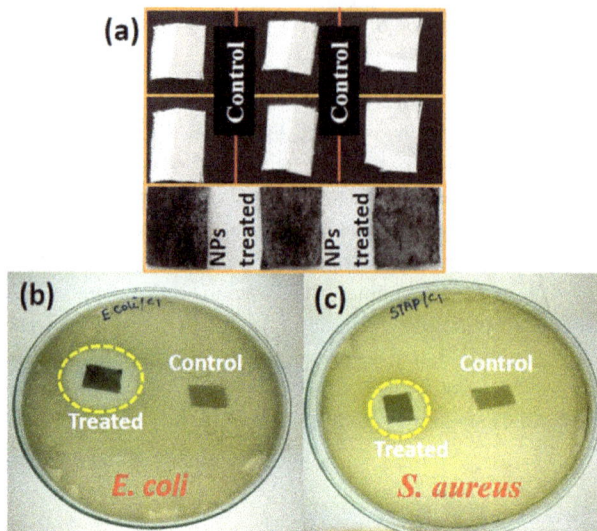

Fig. 6 The zone inhibition for various bacterial strains by FeONPs treated fabrics (a) preparation of fabrics (b) E. coli and (c) S. aureus (Reprinted with permission from Ref. [85]. Copyright 2020, Elsevier)

4.4   Photocatalytic inactivation using nano-flower ZnO catalysts

The synthesized ZnO nanoflowers with 20 ml of Carica papaya milk (CPM) latex is tested for its antibacterial activity toward Gram negative bacteria such as Klebsiella aerogenes (KA) (NCIM-2098), Pseudomonas desmolyticum (PD) (NCIM-5051), and Pseudomonas aeruginosa (PA) (NCIM-2242) and Gram positive bacteria like Staphylococcus aureus (SA) (NCIM-5022) stains. The agar well diffusion approach with the concentration of 200 and 400 µg/ml is used for this testing [86]. The results shows that the ZnO NPs with higher concentration (400 µg/ml) demonstrated the maximum antibacterial activity of 6.62, 9.0, 7.25 and 6.12 mm in diameter, meanwhile, the low concentration of ZnO NPs (200 µg/ml) obtained the minimal activity of 2.38, 6.75, 3.75 and 2.10 mm in diameter for the identical bacterial traces. It is worth noting that the growth of the bacterial is significantly decreased upon increasing the concentration of ZnO NPs. Therefore, the effective antibacterial activity of ZnO nanostructure towards all the bacterial strains has been proven. Most activity is the direct interaction between the ZnO nanoparticle with the outer surface of the membrane of the bacteria, which leading to the disruption of the cellular membrane [87].

In notable, gram-negative microorganism has a lipopolysaccharide which shields the cytoplasmic layer from outside synthetic mixes. The reactive species for example O_2^-, OH^-, H_2O_2, have triggered the damage on the external cell layer of microorganisms [88]. In such as case, the testing has been cultivated inside the dim circumstances. It is assumed that the surface of the ZnO nanostructure have producing the superoxide radicals in which the destruction of cell membranes might possibly accomplised by the reactive species such as OH^-, H_2O_2, and O_2^{2-}.

4.5 TiO₂ nanoparticles to inactivate pathogens in water

TiO_2 is the most regularly utilized nanoparticle to inactivate pathogens in water. It has a wide assortment of business bundles that include: dinners undertaking, photocatalytic media, gas sensor, paint and excellence industry as a white shade, water treatment, air refinement, sun powered power, UV safeguard, semiconductor industry, and rural venture, and so forth [89].

Table 1 *The efficiency of TiO₂ NPs as antimicrobial agent. (Reprinted with permission from Ref. [89], Copyright 2020, Elsevier)*

Elements	Properties	Irradiation process	Matrix	Inactivation effiency	Ref.
TiO₂ powder (Degussa P-25)	Surface area 60 m²/g	Solar light (1660µE s-1 m-2 & dark)	Drinking water	E.coli(99.97% in light after 60 min; 99.96% in dark after 60 min) Initial TiO2 conc.1.0g/L & intial E.coli conc. 106 CFU/ml	[92]
TiO₂-Fe₂O₃ composite	Particle size 10 nm; Surface area 60 m²/g	UV (11 W, Wavelength 253.7 nm)	Drinking water (pH-7.0)	E. coli (99.9% after 1 min.) Initial TiO2 conc. 0.2 g/l & initial E. coli was 107 CFU/100 ml	[93]
TiO₂ powder (Degussa P-25)	Surface area 60 m²/g	UV (18 W, Wavelength 300-420 nm; 7.9 x 10⁻⁶ einsteins/l/s)	Drinking water (pH-7.1)	MS2 Phage, a virus species (88.78% after 120-min) & E. coli (99.43% after 120-min.) Initial TiO2 conc. 1.0 g/l & initial E. coli conc. 6.4 × 105 CFU/ml; MS2 same.	[94]
TiO₂ film	Particle size 8-10 nm; Surface area 147 m²/g	UV (2 x 15 W, Wavelength 365 nm; 3.48mW/cm²) & dark	Wastewater	MC-LR (100% after 150-minute; 24% after 180-minute; E. coli (100% after 120-minute) Initial TiO2 conc. 62.2 µg/cm2 of film & initial E. coli conc. 106-107 CFU/ml)	[95]
TiO₂ & N-doped TiO₂ (Doped by Ethylenediamine & Ethanolamine	Particle size 10 nm; Surface area 80-120 m²/g;	Solar light (10 W/m²)	Synthetic water (pH-7.2)	E. coli (99.99% by TiO2 after 120-min.; 100% by TiO2-N-Ethylenediame after 120-min.; 99.9999% by TiO2–N-Ethanolamine after 120-min.) Initial TiO2 conc. 0.1 g/l & initial E. coli conc. 8.9 × 108 CFU/ml	[96]
TiO₂ NP A12	Particle size 12 nm; Surface area 82 m²/g	Dark	Water	C. metallidurans CH34 (0% after 24-hour at 500 mg/l dose); E. coli (45% after 24-hour at 500 mg/l dose)	[97]
TiO₂ NP A17	Particle size 17 nm; Surface area 122 m²/g			C. metallidurans CH34 (0% after 24-hour at 500 mg/l dose); E. coli (80% after 24-hour at 500 mg/l dose)	
TiO₂ NP A25	Particle size 25 nm; Surtace area 46 m²/g			C. metallidurans CH34 (0% after 24-hour at 500 mg/l dose); E. coli (75% after 24-hour at 500 mg/l dose)	
TiO₂ NP A140	Particle size 142 nm; Surface area 10 m²/g			C. metallidurans CH34 (0% after 24-hour at 500 mg/l dose); E. coli (2% after 24-hour at 500 mg/l dose)	

TiO₂ NP R20	Particle size 21 nm; Surface area 73 m²/g			C. metallidurans CH34 (0% after 24-hour at 500 mg/l dose); E. coli (45% after 24-hour at 500 mg/l dose)	
TiO₂ NP R9	Particle size 9 nm; Surface area 118 m²/g			C. metallidurans CH34 (0% after 24-hour at 500 mg/l dose); E. coli (20% after 24-hour at 500 mg/l dose)	
TiO₂ NP R700	Particle size 707 nm; Surface area 2 m²/g			No inactivation in both case	[97]
TiO₂ powder (Degussa P-25)		UV (4 x 8 W, UV-A; Wavelength 315 – 400 nm; 2.5 mW/cm²)	Water	E. coli (95% after 2 min.) Initial TiO2 conc. 0.1 g/l & initial E. coli conc. 7 × 107 CFU/ml;	[98]
2.46% Ag doped TiO₂	Particle size 10-50 nm			E. coli (99% after 2 min.) Initial TiO2 conc. 0.1 g/l & initial E. coli conc. 7 × 107 CFU/ml;	
4.36% Ag doped TiO₂				E. coli (99.99% after 2 min.) Initial TiO2 conc. 0.1 g/l & initial E. coli conc. 7 × 107 CFU/ml;	
5.95% Ag doped TiO₂				E. coli (99.9999% after 2 min.) Initial TiO2 conc. 0.1 g/l & initial E. coli conc. 7 × 107 CFU/ml;	
0.5% Pt-TiO₂	Particle size 20 nm; Surface area 38 m²/g	150 W Xenon; Flux 17.4 x 10⁻⁸ – 58 x10⁻⁸ einsteins/(Ls);13.1 x 10⁻³ W/cm²	Wastewater	E. coli (99.99% after 1 h) Initial TiO2 conc. 0.1–1.0 g/l & initial E. coli was 1200 CFU/ml	[99]

Markowska-Szczupak et al., have clarified microorganism's inactivation relies upon some of a few components, for examples the centralization of TiO₂, a type of microorganism, power and frequency of gentle, recognition of hydroxylation, pH, and temperature, accessibility of oxygen and receptive oxygen species (ROS)and maintenance time [90]. Li et al., have affirmed the TiO₂ molecule can inactivate microorganisms beneath UV/sun illumination by method for shaping some ROS comprehensive of hydroxyl radical (•OH), superoxide radical ($O_2^{\bullet-}$) and hydrogen peroxide (H₂O₂) [91]. From the writing overview [92-99], the antimicrobial execution of the TiO₂ nanoparticles is expressed in Table 1.

4.6 Photocatalytic inactivation of pathogens in water using extracellular biosynthesis of cobalt ferrite nanoparticles

In the previous work reported by El-Sayed R. El-Sayed and co-workers [100], it is found that the synthesized cobalt ferrite nanoparticles exhibited a broad spectrum of antibacterial activity, in which it prevented the growth of E. coli, Staph. aureus, P. aeruginosa, and K. pneumoniae. Besides that, they obtained results also indicate that the minimum inhibitory concentration (MIC) of the E. coli, Staph. aureus and P. aeruginosa is 250 μg/mL whilst 500 μg/mL for K. pneumoniae. Furthermore, the study on the antifungal activity of the prepared cobalt ferrite nanoparticles toward several plant pathogenic fungi and C. albicans demonstrates that the growths of the tested fungal species have been restrained. Based on the results attained from this work, it reveals that the tested fungal species possesses MIC of 250 μg/mL and the inhibition zones of growth around the agar are measured to be 14.53, 10.53 and 11.76 mm for A. niger, A. solani, and F. oxysporum, respectively. Meanwhile,

Bioinspired Nanomaterials for Energy and Environmental Applications   Materials Research Forum LLC
Materials Research Foundations **121** (2022) 39-82   https://doi.org/10.21741/9781644901830-2

C. Albicans exhibits higher sensitivity towards the nanoparticles in which at a concentration of 250 μg/mL, the inhibition zones of growth around the agar well of 9.53 mm.

4.7   Photocatalytic inactivation using different doses of g-C₃N₄ photocatalysts

Fig. 7(a) shows the photocatalytic purification execution of infections with several dosages of g-C₃N₄ photocatalysts underneath observed light illumination ($\lambda \geq 400$ nm). Based on the obtained results under the dark control, it showed that the concentration of living infections remained unaltered after 360 min, which indicates that there is no harmful consequences of g-C₃N₄ on bacteriophageMS2 infection in the dark (i.e., g-C₃N₄ is biocompatible in the dark). Furthermore, the outcomes of light control test demonstrated that no MS2 have been inactivated without the presence of the g-C₃N₄ photocatalyst, which showed that the inactivation of infections are not caused by the direct photolysis beneath observed visible light irradiation. Though, MS2 can be effectively inactivated with the presence of both g-C₃N₄ and visible light, thus, proving that the virucidal impacts are ascribed to the photocatalysis of g-C₃N₄ semiconductor. It has been observed that the rate of viral inactivation apparently enhanced even with the presence of few amount of g-C₃N₄ (50 mg/L), in comparison with the light control test in the absence of photocatalysts. This observation proved that the efficiency of viral inactivation is significantly affected by the dosage of g-C₃N₄ used. Upon the insertion of photocatalysts, the level of inactivation for MS2 enhanced from ~ 4.5 logs to ~ 6.0 logs when the concentration of photocatalyst increased from 50 mg/L to 100 mg/L and achieved its maximum of ~8 logs at photocatalyst concentration of 150 mg/L under the visible light illumination with the period of 360 minutes. However, the inactivation of MS2 reduced to 7.5 logs when the concentration of g-C₃N₄ photocatalyst is further increased to 200 mg/L. The decrease in the performance of photocatalytic is due to the presence of excessive amount of photocatalysts that cause in diminishing of light penetration. The photocatalytic inactivation of infections is critically influenced by the amount of g-C₃N₄ used; hence the gC3N4 with a concentration of 150 mg/L has been applied throughout the subsequent experiments. [101]. In 360 minutes under the visible light illumination, the 1×10^8 PFU/mL initial concentration of MS2 can be totally inactivated once the g-C₃N₄ with the concentration of 150 mg/L has been loading. The obtained results have been presented in Fig. 7(b).

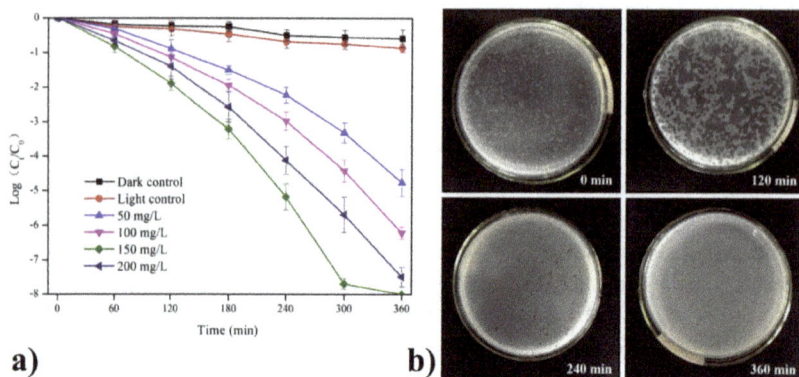

a)   b)

Fig. 7 (a) Photocatalytic inactivation efficiency of MS2 (1 x 10^8 PFU/mL, 100 mL) with the assistance of various doses of g-C_3N_4 under visible light irradiation (b) Photographs of MS2 plaques formation before and after photocatalytic disinfection with g-C_3N_4 (150 mg/L) under visible light irradiation. (Reprinted with permission from Ref. [101], Copyright 2020, Elsevier)

4.8   Photocatalytic disinfection using C. abyssinica tuber extract mediated synthesized ZnO nanoparticles

The studies conducted by researchers have found that zinc oxide nanoparticles (ZnO NPs) have exhibited wide antibacterial activities in opposition to gram negative and positive bacteria [103]. In the work reported by Tura Safawo and co-workers, the prepared ZnO NPs using the C. abyssinica tuber extracts have shown interesting antibacterial activity towards all pathogenic gram negative and gram positive bacteria that have been examined. The agar well diffusion technique has been utilized in order to study the antibacterial activity of prepared ZnO NPs towards the gram positive (Bacillus coagulans, Staphylococcus aureus) and gram negative bacteria (Shigella dysenteriae, S. typhimurium and Sphingomonas paucimobilis) at different concentrations (10, 20, 30 μg/mL). The final outcomes are depicted in Fig. 8 and simplified in Table 2. Among them, Staphylococcus aureus is highly sensitive against ZnO NPs with the inhibition zone of 21 mm whilst the less sensitive tested pathogenic strain is the S. Typhimurium which showed the inhibition zone of 15 mm at 30μg/mL. This suggests that the gram negative bacteria possess high resistivity towards ZnO NPs. The obtained results are in accordance with the one reported by Premanathan et al. [104] and Raghupathi et al. [105]. Likewise, MIC values ZnO NPs towards Staphylococcus aureus is 1.25 μg/mL and 5μg/mL for S. Typhimurium. This uniqueness in affectability among gram negative and positive bacteria is ascribed to their difference in morphology which brings about the distinction in obstruction against the interruption of nanoparticle through the bacteria walls The physiochemical characteristics of NPs including size, shape, chemical composition and solubility are also influencing the efficacy of ZnO Nps towards bacteria [106]. The size of the NPs play an important role in determaining their bioactivity level as the high amount of the NPs can occupy inside the cell membrane and cytoplasmthis if the size is smaller, hence increase its sensitivity [107,108]. The studies conducted by Raghupathi et al. have found that the efficacy in restraining the development of microorganisms increasing when the smaller particles size is used [105]. Besides that, Brayner et al. also found that the rate of bacterial growth can be 100 % been reduced using the prepared ZnO NPs with an average size and concentration of 12 nm and 3 mM, respectively. This is due to the disorder of the E. coli membranes which enhance the permeability of the membrane, thus prompting the accumulation of Nps in the bacterial membrane and cytoplasm regions of the cells [109]. The ZnO NPs which have been synthesized using green technique from C. abyssinica tuber normally possesses an average size of 10.4 nm, in which becomes one of the factors that causing its strong antimicrobial activities.

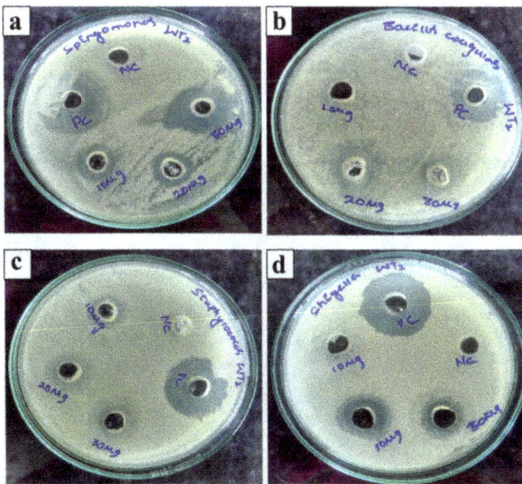

Fig. 8 (a) Sphingomonas paucimobilis, (b) Bacillus coagulans, (c) Staphylococcus aureus and (d) Shigella dysenteriae Plate displaying the zone of inhibition at different concentration of ZnO NPs (Reprinted with permission from Ref. [102], Copyright 2020, Elsevier)

Table. 2 Zone of inhibition of ZnO Nps against bacterial strains (Reprinted with permission from Ref. [102], Copyright 2020, and Elsevier)

Bacteria strains	ZnO NPs concentration			Control (Chromaphenicol) 30 µg/mL
	10 µg/mL	20 µg/mL	30 µg/mL	
Bacilus coagulans	-	17	20	23
Staphylococcus aureus	16	17	21	27
Sphingomonas paucimobilis	11	15	18	24
S. typhimurium	13	9	15	19
Shigella dysenteriae	10.5	15	16	22

4.9   Photocatalytic inactivation analysis using Biosynthesis of Ag deposited phosphorus and sulfur co-doped g-C₃N₄ (Ag-PSCN) nanocomposite

An experiment has been conducted under the dark and visible light conditions in order to investigate the inactivation properties of the prepared composites towards E. Coli [110]. Under the darkish circumstance, the detection of viable E. coli cells shows an increasing trends from 0.25 logs to 0.45 logs of decreases when the mass percentages of Ag to phosphorus and sulfur co-doped g-C$_3$N$_4$ (PSCN) is increased from 1 % to 8 %. It is interesting to note that both g-C3N4 (CN) and PSCN have exhibited insignificant cytotoxicity towards E. coli under the dark circumstance while the composites of Ag/PSCN displayed barely inactivation efficiencies due to the low content of released Ag$^+$ [111].

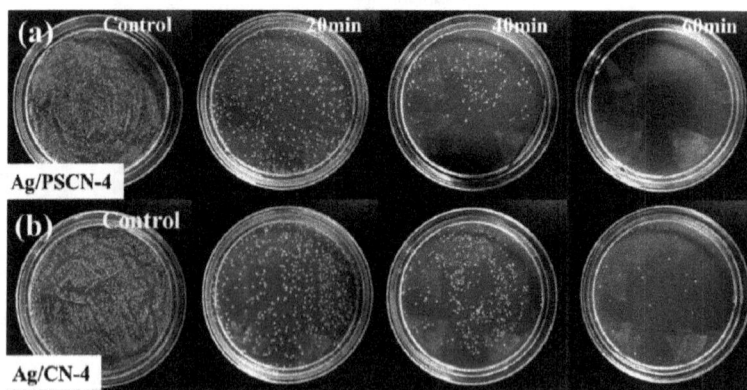

Fig. 9 E. coli colonies growth in the presence of Ag/PSCN-4 (a) and Ag/CN-4 (b) during the photocatalytic inactivation process (Reprinted with permission from Ref. [110], Copyright 2020, Elsevier).

Bioinspired Nanomaterials for Energy and Environmental Applications   Materials Research Forum LLC
Materials Research Foundations **121** (2022) 39-82   https://doi.org/10.21741/9781644901830-2

The results of inactivation of the prepared composites towards E. coli obtained under visible light illumination are presented in Fig. 9. It is observed that only 0.6 and 1.1 logs of cells are inactivated by CN and PSCN, respectively, over the period of 60 minutes under the illumination of visible light. On the contrary, all the Ag/PSCN composites show an improvement in the inactivation activities, in opposition to PSCN. The inactivation efficiencies have shown an increasing trend when the content of deposited Ag rose from 1 % to 4 %. However, it displayed a decreasing trend upon increasing the content of Ag to 8 %. Among the composites, Ag/PSCN-4 have exhibits the highest bactericidal efficiency in which it can totally inactivated 7.0 logs of E. coli after being irradiated by visible light for 60 minutes. It is no doubt that this phenomenon proved that there are a few factors have played a dominant roles in increasing the efficiency of the photocatalytic inactivation such as the tendency of E. coli to attached on the surface of Ag/PSCN, the interaction between Ag and PSCN, and the resulting photo-induced active species [112,113]. The diminished inactivation ability of Ag/PSCN-8 is suspected due to the presence of abundance deposited Ag particles that served as the recombination centers, thus caused the decreasing in the charge separation efficiency as well as the production of active species [114]. **Fig. 10** demonstrates the possible mechanism for inactivation of E. coli using Ag/ PSCN-4.

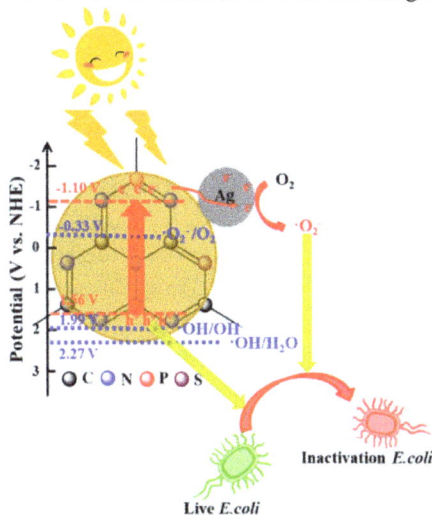

Fig. 10 Photocatalytic inactivation mechanism of Ag/PSCN-4 under visible light irradiation. (Reprinted with permission from Ref. [110], Copyright 2020, and Elsevier)

4.10 Photocatalytic disinfection of pathogens in water using Nd_2WO_6/ZnO incorporated on GO (NWZG) nanocomposite

The antibacterial action of the NWZG nanocomposite is analyzed against E. coli (Gram-poor) and B. cereus (Grampositive) microscopic organisms by utilizing the agar circle dispersion system. Based on the obtained results, the NWZG composite with the concentration of a 100μg/μL exhibits the higher antibacterial activity towards E. coli in contrast to the B. cereus and the 100 μg/μL concentration of chloramphenicol. The standard chloramphenico demonstrates inhibition zone value of 16 mm and 18 mm for gram negative E. coli bacteria and gram positive B. cereus bacteria, respectively. In comparison, the NWZG nanocomposite with the concentration of50 μg/μL obtained an inhibition zone of 15 mm and 8 mm for E. coli and B. cereus bacteria, respectively, and it increased to 18 mm for E. coli and 11 mm for B. cereus bacteria upon increasing the concentration of NWZG nanocomposite to 100μg/μL. Generally, there are few factors that affecting the efficacy of antibacterial such as the particle size and morphology, and etc. [115]. It is believed that the NWZG nanocomposite is breaking the wall of the cell due to the strong electrostatic interaction between the cell wall and the nanocomposite. This phenomenon plays an important part in the activity study [115–118]. Results from these electrostatic interactions will induce the generation of ROS in which hinder the growth of bacterial. From the NWZG system, by the illumination of light, the electron become excited and jumped from valence band to conduction band. This energetic electron at the conduction band will interact with oxygen molecule to form superoxide anion. [119]. This superoxide anion will further interacts with water molecules to produce H_2O_2 that will be form inside the nanocomposite surface and next crushes the bacterial cell membrane. At this stage, in the VB the electrons from water to deliver hydroxyl radicals. The incredible H_2O_2 has demolished the microscopic organisms, however the hydroxyl radical and superoxide anion are unable to infiltrate inside the cell film and in this way it self-destructs the proteins and lipids inside the external surface of microbes. Conclusively, the prepared NWZG nanocomposite shows a great potential as the antibacterial agent that capable to demolish the organisms in water.

4.11 Visible-light-driven photocatalytic disinfection of bacteria's using Zea mays L. dry husk mediated bio-synthesized copper oxide nanoparticles

A study on the antibacterial effectiveness of CuO NPs against Escherichia coli 518133, Staphylococcus aureus 9144, Pseudomonas aeruginosa and Bacillus licheniformis has been performed and the results obtained shows that the diameters of the inhibition zones varied in the range of 8 mm to 15 mm, depending on the annealing temperatures used in the synthesis process. The CuO NPs produced under the annealing temperature of 300 °C

(CuO_300) demonstrated the largest inhibition zone diameter (15 mm) compared to other CuO NPs type. Meanwhile, Cu_2O and CuO_600 have an inhibition area diameter of 10mm and 7 mm, respectively. It is worth noting that CuO_300 and Cu_2O are highly effective to prevent the growth of Escherichia coli 518,133 and Staphylococcus aureus 9144. Cu_2O had shown a significant effect against both Pseudomonas aeruginosa and Bacillus licheniformis in comparison to other two types of CuO. It has been suggested to use CuO_300 for preventing the growth of Escherichia coli 518,133 and Staphylococcus aureus 9144, whilst Cu_2O for inhibit the growth of Pseudomonas aeruginosa and Bacillus licheniformis. [120].

4.12 Efficient photocatalytic disinfection of Escherichia coli O157:H7 using C70-TiO₂ hybrid material under visible light irradiation

In the technique of photocatalytic disinfection, the bacterial cellular walls are damaged first, suggesting that the fluorescent staining technique is an instantaneous and accurate approach, and accordingly it changed into used within the following experiment.

4.12.1 Photocatalytic disinfection mechanism

It is nice that usually valence band (VB) holes and conduction band (CB) electrons may be generated whilst the photocatalyst is irradiated via incident light with a right wavelength [121]. VB holes can oxidize the H_2O to provide •OH and the photogenerated CB electrons can reduce the floor absorbed O_2 to offer the superoxide radical anion O_2-•. The produced •OH and O_2-• radicals in addition to VB holes can disinfect the microorganism. But, there's nonetheless a heated argument about which reactive species performs a greater notable feature within the photocatalytic disinfection machine [122, 123]. Consequently, to address this issue, we used one of kind scavengers in my view to do away with the appropriate reactive species. Isopropanol and sodium oxalate had been used as the scavengers for •OH and VB holes, respectively. The isopropanol and sodium oxalate can't disinfect the E. coli O157:H7 inside the absence of photocatalyst. Within the presence of sodium oxalate, the photocatalytic disinfection impact of C70-TiO₂ is sort of similar to that without the scavenger, indicating that VB holes have almost no motion in the disinfection. Apparently, with the addition of isopropanol to eliminate •OH in bulk, the photocatalytic disinfection effect of C70-TiO₂ is truly inhibited (the survival price of E. coli O157:H7 rose from 25% to 74% after 2 h irradiation), indicating that •OH generated through the photocatalyst performs a critical characteristic in the disinfection of E. coli O157:H7.

4.13 Antibacterial assay on SnO_2 doped GO and CNT under visible light irradiation

It is pleasantly regular that valence band (VB) gaps and conduction band (CB) electrons might be produced while the photocatalyst is illuminated through occurrence light with a correct frequency [121]. VB openings can oxidize the H_2O to give •OH and the photogenerated CB electrons can decrease the floor consumed O_2 to offer the superoxide radical anion O_2-•. The delivered •OH and O_2-• radicals notwithstanding VB gaps can sanitize the small scale living being. Be that as it may, there is, in any case, a warmed contention about which responsive species plays out a more prominent outstanding element inside the photocatalytic purification machine [122, 123]. Thusly, to address this issue, we utilized one of the kind scroungers in my view to get rid of the fitting receptive species. Isopropanol and sodium oxalate had been utilized as the scroungers for •OH and VB openings, individually. The isopropanol and sodium oxalate can't clean the E. coli O157:H7 inside the nonappearance of the photocatalyst. Inside the nearness of sodium oxalate, the photocatalytic purification effect of $C70-TiO_2$ is kind of like that without the forager, showing that VB openings have basically no movement in the sanitization. Clearly, with the expansion of isopropanol to wipe out •OH in mass, the photocatalytic cleansing impact of $C70-TiO_2$ is truly repressed (the endurance cost of E. coli O157:H7 rose from 25% to 74% after 2 h light), showing that •OH produced through the photocatalyst plays out a basic trademark in the cleansing of E. coli O157:H7. **Fig. 11** represents the photocatalytic antibacterial activity on (a) E. coli, (b) P. gramin is treated with the positive control and SnO_2 doped GO and CNT nanocomposite. The comparisons of the SnO2 photocatalysts against Bacterial species are portrayed in **Table 3.**

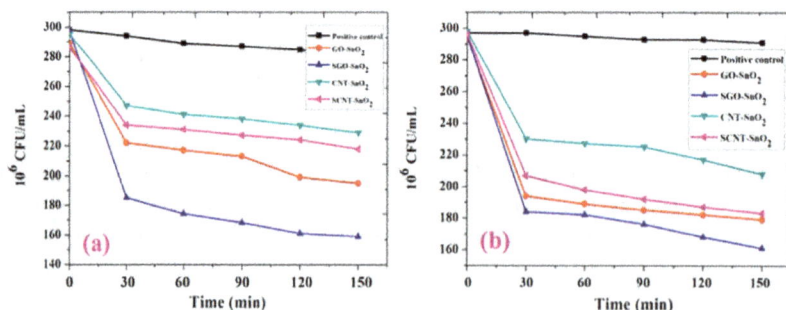

Fig. 11 Photocatalytic antibacterial activity on (a) E. coli, (b) P. graminis treated with the positive control and nanocomposite. (Reprinted with permission from Ref. [124], Copyright 2020, Elsevier)

Table. **3** *Comparison of the SnO₂ photocatalysts against Bacterial species (Reprinted with permission from Ref. [124], Copyright 2020, Elsevier).*

No.	Photocatalyst	Loading Concentration	Light used	Bacteria Used	CFU/mL	Zone of Inhibition (mm)	Ref.
1	SnO₂	0.03 mol	Dark/Visible	E.coli	$35/27 \times 10^8$	NA	[127]
2	SnS₂	0.005 mol	Dark/Visible	E.coli	$27/20 \times 10^8$	NA	
3	SnO₂/ SnS₂	0.03/0.005 mol	Dark/Visible	E.coli	$19/13 \times 10^8$	NA	
4	SnO₂	0.03 mol	Dark/Visible	S. aureus	$35/27 \times 10^8$	NA	
5	SnS₂	0.005 mol	Dark/Visible	S. aureus	$27/21 \times 10^8$	NA	
6	SnO₂/ SnS₂	0.03/0.005 mol	Dark/Visible	S. aureus	$22/15 \times 10^8$	NA	
7	SnO₂/ ZnOₓ	300 mg	Visible	E.coli	NA	25	[128]
8	SnO₂/ ZnOₓ	300 mg	Visible	S. aureus	NA	24	
9	Co-SnO₂	0.011 mol	NA	E.coli	NA	15	[129]
10	Co-SnO₂	0.009 mol	NA	B. subtilts	NA	15	
11	TiO₂-SnO₂	5 mol%	UV	E.coli	47%	NA	[130]
12	SnO₂	3.33 mg/mL	UV	S. aureus	58%	NA	[131]
13	SnO₂	3.33 mg/mL	UV	P.aeruginosa	65%	NA	
14	TiO₂-SnO₂-Fe³⁺	1:3:0.5 ratio	UV	E.coli	NA	10	[132]
15	TiO₂-SnO₂-Fe³⁺	1:3:0.5 ratio	UV	S.typht	NA	15	
16	TiO₂-SnO₂-Fe³⁺	1:3:0.5 ratio	UV	S. aureus	NA	10	
17	TiO₂-SnO₂-Fe³⁺	1:3:0.5 ratio	UV	L. monocytogenes	NA	14	
18	GO-SnO₂	200 µg/ml	Visible	E.coli	1.95×10^8	18	[This work]
19	SGO-SnO₂	200 µg/ml	Visible	E.coli	1.59×10^8	21	
20	CNT-SnO₂	200 µg/ml	Visible	E.coli	2.29×10^8	9	
21	SCNT-SnO₂	200 µg/ml	Visible	E.coli	2.18×10^8	14	
22	GO-SnO₂	200 µg/ml	Visible	P.gramints	1.79×10^8	12	
23	SGO-SnO₂	200 µg/ml	Visible	P.gramints	1.61×10^8	12	
24	CNT-SnO₂	200 µg/ml	Visible	P.gramints	2.08×10^8	7	
25	SCNT-SnO₂	200 µg/ml	Visible	P.gramints	1.83×10^8	8	

4.14   Photocatalytic inactivation on ZnO nanoparticle under visible light irradiation

The impact of comparable concentrations of synthesized ZnO morphologies against S. pnemoniae, S. aureus, and E. coli and E. hermannii have been studied the usage of properly diffusion approach as a cause of ZnO absorption [133]. The antibacterial inhibition development of gram superb microorganism and gram negative bacteria changed into planned by using the use of ZnO nanoparticles and Nitrofurantoin changed into used as the standard. The inhibitory result of ZnO nanoparticles and the standard Nitrofurantoin are plotted in Fig. 12.

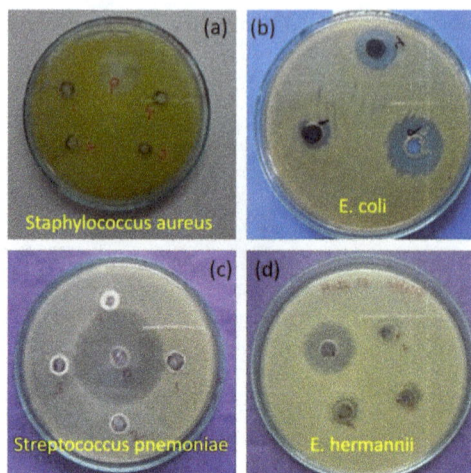

Fig. 12 Inhibitory effect of ZnO thin film on bacteria's in composite structure (a-d), (Reprinted with permission from Ref. [133] Copyright 2020, Elsevier)

A ZnO nanoparticle uncovers the better restraint inside the expansion against a board of small-scale life forms. Among these S. aurei become resolved to be touchy to the ZnO nanoparticles at the consciousness of 200µg/ml of example with an area of the hindrance of 15 mm. The ZnO nanoparticles showed astonishing diversion against E. hermannii with a district of restraint width of 16 mm at 400µg/ml. While nanoparticles show significantly less enthusiasm inside the development hindrance and antibacterial enthusiasm toward the E. coli and S. pneumonia microscopic organisms at all of the consideration of nanoparticles. The no-doubt move for the improvement of restraint conclusive outcomes of miniaturized scale living being by methods for the nanoparticles can be pondered in two grouped manners. Directly here, in the primary case Zn^{2+} from the nanoparticles will sidestep inside the bacterial cell and relates with the horrible segment of the microscopic organisms and principals to the improvement restraint and convincingly bacterial cell kicks the bucket. Inside the 2d case light power got enlightened at the nanoparticles wherein the photon quality is more than the bandgap, the better force electrons (e^-) from the valence band are eager to the conduction band, leaves the fine opening (h^+) in the valence band, and results inside the assembling of responsive oxygen species (ROS) like superoxide anion (O_2^{-}) by means of reductive technique and generally receptive hydroxyl radicals (OH^-) by means of oxidative strategy. The superoxide anion responds with H_2O atoms to give hydrogen peroxide made on the shallow of the nanoparticles and has style to abrogate the bacterial cell film (Fig. 13). Accordingly, those results are indicated that the consideration of nanoparticles in nano-degree has a higher antibacterial action and might be utilized inside the annihilation of organisms.

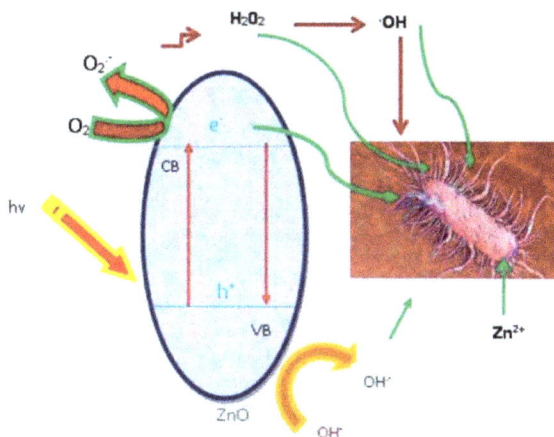

Fig. 13 Possible mechanism of photoexcitation in composite structure (Reprinted with permission from Ref. [133] Copyright 2020, Elsevier)

4.15   Photocatalytic disinfection of bacterial pathogens using MgO nanostructures

In recently, MgO nanostructures are examined in opposition to styles of microorganism like gram-negative bacteria (G -): Escherichia coli, Shigella flexneri, Salmonella typhi, Proteus mirabilis, Vibrio cholera, and Aeromonas hydrophila and gram-positive (G +): Bacillus subtilis and Rhodococcus rhodochrous following the disc diffusion technique (Fig. 14) to find out the antibacterial activity similar to the usual antibiotics (Ampicillin). The ZOI of MgO nanostructures is presented in Table 4. The bactericidal activity of MgO nanostructures in large element depends at the ROS this is particularly assigned to a higher precise surface place, polar surface, crystal period, morphology, a lift in oxygen vacancies, the diffusion capacity of the chemical molecule and additionally the discharge of Mg^{2+}. The bactericidal mechanism of MgO consists of sterilization by way of the use of particle exudation, absorption, and complicated feature. This takes a take a look at has discovered the relationship amongst nanostructures and bactericidal activity [134].

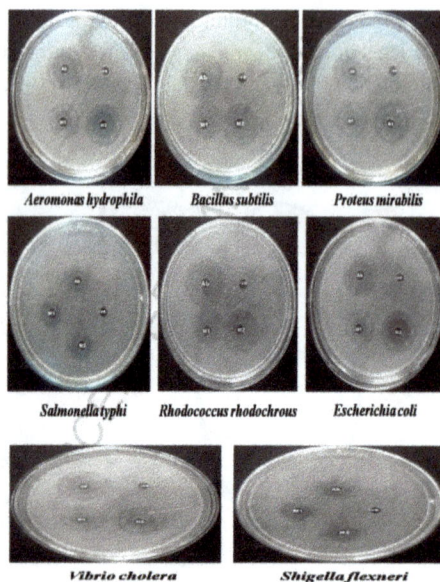

Fig. 14 Antibacterial activity of MgO nanostructures against human bacterial pathogens (M 1: MgO nanoparticles prepared by hydrothermal method and M2: MgO nanorods prepared by microwave assisted method) (Reprinted with permission from Ref. [134] Copyright 2020, Elsevier)

Table. 4 Antimicrobial activity of MgO nanostructures against bacterial pathogenic organisms, mean zone of inhibition (mm) (Reprinted with permission from Ref. [134], Copyright 2020, Elsevier).

Bacterial Strains	Gram reaction	Zone of Inhibition (mm)		Positive control	Negative control
		M 1	M 2		
Shigella flexneri	-ve	14	17	14	-
Escherichia coli	-ve	20	24	30	-
Proteus mirabilis	-ve	22	24	27	-

Aeromonas hydrophila	-ve	24	25	27	-
Bacillus subtilis	+ve	23	27	30	-
Vibrio cholera	-ve	21	23	26	-
Rhodococcus rhodochrous	+ve	24	27	30	-
Salmonella typhi	-ve	13	16	17	-

In general, smaller crystallite size with a high surface area brings about an over the top antibacterial activity. The superoxide radicals, hydrogen peroxide responds to the ROS, hurting the deoxyribonucleic acid and cell proteins which bring about portable demise. Fig.14 demonstrates the size of the zone of hindrance normal around each plate stacked with MgO nanoparticles; the MgO nanorods show the higher antibacterial action. The antimicrobial activity of the MgO nanostructures is clarified as follows. The instrument of light-initiated innovation of ROS is given by utilizing conditions:

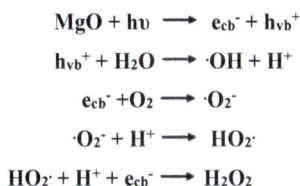

$$MgO + h\upsilon \longrightarrow e_{cb}^- + h_{vb}^+$$
$$h_{vb}^+ + H_2O \longrightarrow \cdot OH + H^+$$
$$e_{cb}^- + O_2 \longrightarrow \cdot O_2^-$$
$$\cdot O_2^- + H^+ \longrightarrow HO_2\cdot$$
$$HO_2\cdot + H^+ + e_{cb}^- \longrightarrow H_2O_2$$

As noted earlier, the photocatalysis technique appears to be the foremost crucial mechanism for an antibacterial activity. Mainly, ROS created at the bottom of the nanoparticles within the presence of light reasons aerophilous strain on the microorganism cells and in the long run bring about demise. ROS incorporates the reactive hydroxyl radical (OH), plenty less toxic superoxide anion radical (O_2^-), and a inclined oxidizer hydrogen peroxide (H_2O_2). Superoxide anion radical (O_2^-) has successively reacted with H^+ to provide (HO_2^-), upon the following collisions with electrons to provide you with hydrogen peroxide anions (HO_2^-). They react with hydrogen ions to provide molecules of H_2O_2. The generated (H_2O_2) penetrates to the deoxyribonucleic acid, cell membranes, and cellular proteins for destroying the microbes. The larger ROS is predicated upon upon the smaller crystallite duration (20 nm (M 1) & 14 nm (M2)) with the excessive particular area and conjoint boom inside the surface defects.

The second one purpose for the bacterial activity is completed through as quickly because the Mg^{2+} ions are discharged from the MgO surface, it comes into touch with the cell membranes of the microbes. The antibacterial activity research advises that the synthesized nanomaterials act right now at the pathogenic bacteria to damage the cells and membrane integrity which ends up within the dying of the bacterial pathogens. The MgO nanorods showcase comparable antibacterial activity in opposition to all of the examined pathogens. The zones of inhibition of the tested pathogens are given in Table 4. In correspondence with current results, it's far claimed that the higher antibacterial activity is derived from the powerful surface of the MgO because of a particular ground place, morphology, and smaller crystallite duration [135, 136].

Conclusions

This chapter provides a focus on latest research activities that utility of bio-stimulated photocatalysts for pathogens disinfection underneath the illumination of light energy. The photo-excitation of a few semiconductors results in the production of ROS that may inactivate microorganisms and the bacterial disinfection of different nano-dependent photocatalysts along with TiO_2, ZnO, Fe_2O_3, NiO and metal sulphides, etc., are defined. This chapter concludes the bio-inspired metal oxide photocatalysts have unique benefits of earth abundance, very low price, eco-friendliness, easy structure, and easy to synthesis and special interest is given to the development of rising areas, along with environmental in addition to strength substances. The developments of an easy, yet less expensive water disinfection era to assist cope with waterborne ailment chance inside the growing regions. The bacterial pathogens inactivation mechanisms are also explained in the chapter. The development of novel bio-stimulated photocatalytic substances to be used in disinfection packages is explained. Finally, it is concluded with a discussion about the development of bio-inspired metallic oxide photocatalysts for bacterial disinfection under light energy. Among the bio-inspired materials were tested for their disinfection properties. Photocatalytic disinfection manner is a developing, challenging, and multi-disciplinary field that requires the cooperation of researchers in microbiology and the physical sciences.

References

[1]   Dujaradin E, Mann S, Bioinspired materials chemistry, Adv. Mater.,14 (2002) 1-13. https://doi.org/10.1002/1521-4095(20020104)14:1<13::AID-ADMA13>3.0.CO;2-W

[2]   Bhattacharya D, Gupta RK. Nanotechnology and potential of microorganisms, Crit. Rev. Biotechnol., 25 (2005) 199-204. https://doi.org/10.1080/07388550500361994

[3]　Mandal D, Bolander ME, Mukhopadhyay D, Sarkar G, Mukherjee P, The use of microorganisms for the formation of metal nanoparticles and their application, Appl. Microbiol. Biotechnol. 69 (2006) 485-92. https://doi.org/10.1007/s00253-005-0179-3

[4]　Thakkar KN, Mhatre SS, Parikh RY, Biological synthesis of metallic nanoparticles, Nanomed: Nanotechnol. Biol. Med. 6 (2010) 257-62. https://doi.org/10.1016/j.nano.2009.07.002

[5]　Gade A, Ingle A, Whiteley C, Rai M, Mycogenic metal nanoparticles: progress and applications, Biotechnol. Lett. 32 (2010) 593-600. https://doi.org/10.1007/s10529-009-0197-9

[6]　Sharma VK, Yngard RA, Lin Y, Silver nanoparticles: Green synthesis and their antimicrobial activities, Adv. Coll. Interf. Sci. 145 (2009) 83-96. https://doi.org/10.1016/j.cis.2008.09.002

[7]　Vijayaraghavan K, Nalini SP, Biotemplates in the green synthesis of silver nanoparticles, Biotechnol. J., 5 (2010) 109-110. https://doi.org/10.1002/biot.201000167

[8]　Tolaymat TM, El Badawy AM, Genaidy A, Scheckel KG, Luxton TP, Suidan M, An evidence-based environmental perspective of manufactured silver nanoparticles in syntheses and applications: A systematic review and critical appraisal of peer reviewed scientific paper, Sci. Total Environ. 408 (2010) 999-1006. https://doi.org/10.1016/j.scitotenv.2009.11.003

[9]　Rai M, Yadav A, Gade A, Silver nanoparticles as a new generation of antimicrobials, Biotechnol. Adv. 27 (2009) 76-83. https://doi.org/10.1016/j.biotechadv.2008.09.002

[10]　Zhang X, Yan S, Tyagi RD, Surampalli RY, Synthesis of nanoparticles by microorganisms and their application in enhancing microbiological reaction rates, Chemosphere 82 (2011) 489-94. https://doi.org/10.1016/j.chemosphere.2010.10.023

[11]　P. Mohammadzadeh Pakdel, S.J. Peighambardoust, Review on recent progress in chitosan-based hydrogels for wastewater treatment application, Carbohydr. Polym., 201 (2018) 264-279. https://doi.org/10.1016/j.carbpol.2018.08.070

[12]　C. Zhang, Y. Li, D. Shuai, Y. Shen, D. Wang, Progress and challenges in photocatalytic disinfection of waterborne Viruses: A review to fill current knowledge gaps, Chem. Eng. J., 355 (2019) 399-415. https://doi.org/10.1016/j.cej.2018.08.158

[13]　T. Tatarchuk, N. Paliychuk, R.B. Bitra, A. Shyichuk, Mu. Naushad, I. Mironyuk, D. Ziółkowska, Adsorptive removal of toxic methylene blue and acid orange 7 dyes from aqueous medium using cobalt-zinc ferrite nanoadsorbents, Desalin. Water Treat., 150 (2019) 374-385. https://doi.org/10.5004/dwt.2019.23751

[14]   T. Cai, Y. Liu, L. Wang, W. Dong, G. Zeng, Recent advances in round-the-clock photocatalytic system: Mechanisms, characterization techniques and applications, J. Photochem. Photobiol. C, 39 (2019) 58-75. https://doi.org/10.1016/j.jphotochemrev.2019.03.002

[15]   A. Kumar, G. Sharma, M. Naushad, A.H. Al-Muhtaseb, A. Kumar, I. Hira, T. Ahamad, A.A. Ghfar, F.J. Stadler, Visible photodegradation of ibuprofen and 2,4-D in simulated waste water using sustainable metal free-hybrids based on carbon nitride and biochar, J. Environ. Manage., 231 (2019) 1164-1175. https://doi.org/10.1016/j.jenvman.2018.11.015

[16]   R. Djellabi, B. Yang, Y. Wang, X. Cui, X. Zhao, Carbonaceous biomass-titania composites with TiOC bonding bridge for efficient photocatalytic reduction of Cr(VI) under narrow visible light, Chem. Eng. J., 366 (2019) 172- 180. https://doi.org/10.1016/j.cej.2019.02.035

[17]   A. Kumar, G. Sharma, M. Naushad, T. Ahamad, R.C. Veses, F.J. Stadler, Highly visible active $Ag_2CrO_4/Ag/BiFeO_3@RGO$ nano-junction for photoreduction of CO_2 and photocatalytic removal of ciprofloxacin and bromate ions: The triggering effect of Ag and RGO, Chem. Eng. J., 370 (2019) 148-165. https://doi.org/10.1016/j.cej.2019.03.196

[18]   S.T. Akar, T. Akar, Z. Kaynak, B. Anilan, A. Cabuk, Ö. Tabak, T.A. Demir, T. Gedikbey, Removal of copper (II) ions from synthetic solution and real wastewater by the combined action of dried Trametes versicolor cells and montmorillonite, Hydrometallurgy 97 (2009) 98-104. https://doi.org/10.1016/j.hydromet.2009.01.009

[19]   A. Kumar, C. Guo, G. Sharma, D. Pathania, M. Naushad, S. Kalia, P. Dhiman, Magnetically recoverable ZrO_2/Fe_3O_4/chitosan nanomaterials for enhanced sunlight driven photoreduction of carcinogenic Cr(vi) and dechlorination & mineralization of 4-chlorophenol from simulated waste water, RSC Advances 6 (2016) 13251-13263. https://doi.org/10.1039/C5RA23372K

[20]   M.M. Khin, A.S. Nair, V.J. Babu, R. Murugan, S. Ramakrishna, A review on nanomaterials for environmental remediation, Energy Environ. Sci., 5 (2012) 8075-8109. https://doi.org/10.1039/c2ee21818f

[21]   P. Dhiman, J. Chand, A. Kumar, R.K. Kotnala, K.M. Batoo, M. Singh, Synthesis and characterization of novel Fe@ZnO nanosystem, J. Alloys Compd., 578 (2013) 235-241. https://doi.org/10.1016/j.jallcom.2013.05.015

[22]   S. Garcia-Segura, E. Brillas, Applied photoelectrocatalysis on the degradation of organic pollutants in wastewaters, J. Photochem. Photobiol. C, 31 (2017) 1-35. https://doi.org/10.1016/j.jphotochemrev.2017.01.005

[23]   Z. He, X. Cheng, G.Z. Kyzas, J. Fu, Pharmaceuticals pollution of aquaculture and its management in China, J. Mol. Liq. 223 (2016) 781-789. https://doi.org/10.1016/j.molliq.2016.09.005

[24]   R. Mirzaei, A. Mesdaghinia, S.S. Hoseini, M. Yunesian, Antibiotics in urban wastewater and rivers of Tehran, Iran: Consumption, mass load, occurrence, and ecological risk, Chemosphere 221 (2019) 55-66. https://doi.org/10.1016/j.chemosphere.2018.12.187

[25]   A. Kumar, A. Kumar, G. Sharma, M. Naushad, F.J. Stadler, A.A. Ghfar, P. Dhiman, R.V. Saini, Sustainable nanohybrids of magnetic biochar supported g-C3N4/FeVO4 for solar powered degradation of noxious pollutants- Synergism of adsorption, photocatalysis & photo-ozonation, J. Clean. Prod. 165 (2017) 431-451. https://doi.org/10.1016/j.jclepro.2017.07.117

[26]   S. Kant, D. Pathania, P. Singh, P. Dhiman, A. Kumar, Removal of malachite green and methylene blue by Fe0.01Ni0.01Zn0.98O/polyacrylamide nanocomposite using coupled adsorption and photocatalysis, Appl. Catal. B, 147 (2014) 340-352. https://doi.org/10.1016/j.apcatb.2013.09.001

[27]   S. Sun, R. Zhao, Y. Xie, Y. Liu, Photocatalytic degradation of aflatoxin B1 by activated carbon supported TiO2 catalyst, Food Control, 100 (2019) 183-188. https://doi.org/10.1016/j.foodcont.2019.01.014

[28]   Constantin, C., Neagu, M., Bio-inspired nanomaterials - a better option for nanomedicine, Trends Toxicol. Related Sci., 1 (2017) 2-20. https://doi.org/10.1016/j.carbpol.2018.02.012

[29]   S.M. Noorbakhsh-Soltani, M.M. Zerafat, S. Sabbaghi, A comparative study of gelatin and starch-based nanocomposite films modified by nano-cellulose and chitosan for food packaging applications, Carbohydrate Polymers 189 (2018) 48-55.

[30]   Y. Mu, Y. Fu, J. Li, X. Yu, Y. Li, Y. Wang, X. Wu, K. Zhang, M. Kong, C. Feng, X. Chen, Multifunctional quercetin conjugated chitosan nano-micelles with P-gp inhibition and permeation enhancement of anticancer drug, Carbohydrate Polymers 203 (2019) 10-18. https://doi.org/10.1016/j.carbpol.2018.09.020

[31]   F. Hassan Hassan Abdellatif, J. Babin, C. Arnal-Herault, L. David, A. Jonquieres, Grafting cellulose acetate with ionic liquids for biofuel purification membranes : Influence of the anion, Carbohydrate Polymers 196 (2018) 176-186. https://doi.org/10.1016/j.carbpol.2018.05.008

[32]   A. Tabriz, M.A. Ur Rehman Alvi, M.B. Khan Niazi, M. Batool, M.F. Bhatti, A.L. Khan, A.U. Khan, T. Jamil, N.M. Ahmad, Quaternized trimethyl functionalized chitosan

based antifungal membranes for drinking water treatment, Carbohydrate Polymers 207 (2019) 17-25. https://doi.org/10.1016/j.carbpol.2018.11.066

[33]   K. Zhang, P. Tao, Y. Zhang, X. Liao, S. Nie, Highly thermal conductivity of CNF/AlN hybrid films for thermal management of flexible energy storage devices, Carbohydrate Polymers 213 (2019) 228-235. https://doi.org/10.1016/j.carbpol.2019.02.087

[34]   H. Du, W. Liu, M. Zhang, C. Si, X. Zhang, B. Li, Cellulose nanocrystals and cellulose nanofibrils based hydrogels for biomedical applications, Carbohydrate Polymers 209 (2019) 130-144. https://doi.org/10.1016/j.carbpol.2019.01.020

[35]   A.H. Jawad, M.A. Nawi, Characterizations of the Photocatalytically-Oxidized Cross-Linked Chitosan-Glutaraldehyde and its Application as a Sub-Layer in the TiO2/CS-GLA Bilayer Photocatalyst System, Journal of Polymers and the Environment 20 (2012) 817-829. https://doi.org/10.1007/s10924-012-0434-5

[36]   S. Mallakpour, V. Behranvand, F. Mallakpour, Synthesis of alginate/carbon nanotube/carbon dot/fluoroapatite/TiO2 beads for dye photocatalytic degradation under ultraviolet light, Carbohydrate Polymers (2019) 115138. https://doi.org/10.1016/j.carbpol.2019.115138

[37]   M. Malakootian, A. Nasiri, A. Asadipour, E. Kargar, Facile and green synthesis of ZnFe2O4@CMC as a new magnetic nanophotocatalyst for ciprofloxacin degradation from aqueous media, Process Safety and Environmental Protection 129 (2019) 138-151. https://doi.org/10.1016/j.psep.2019.06.022

[38]   H. Wang, L. Wang, S. Ye, X. Song, Construction of Bi2WO6–TiO2/starch nanocomposite films for visible-light catalytic degradation of ethylene, Food Hydrocolloids 88 (2019) 92-100. https://doi.org/10.1016/j.foodhyd.2018.09.021

[39]   P. Sirajudheen, S. Meenakshi, Facile synthesis of chitosan-La3+-graphite composite and its influence in photocatalytic degradation of methylene blue, International Journal of Biological Macromolecules 133 (2019) 253-261. https://doi.org/10.1016/j.ijbiomac.2019.04.073

[40]   I. Dalponte, B.C. de Sousa, A.L. Mathias, R.M.M. Jorge, Formulation and optimization of a novel TiO2/calcium alginate floating photocatalyst, International Journal of Biological Macromolecules 137 (2019) 992-1001. https://doi.org/10.1016/j.ijbiomac.2019.07.020

[41]   L. Midya, A.S. Patra, C. Banerjee, A.B. Panda, S. Pal, Novel nanocomposite derived from ZnO/CdS QDs embedded crosslinked chitosan: An efficient photocatalyst and effective antibacterial agent, Journal of Hazardous Materials 369 (2019) 398-407. https://doi.org/10.1016/j.jhazmat.2019.02.022

[42]   D.R. Perinelli, L. Fagioli, R. Campana, J.K.W. Lam, W. Baffone, G.F. Palmieri, L. Casettari, G. Bonacucina, Chitosan-based nanosystems and their exploited antimicrobial activity, European Journal of Pharmaceutical Sciences 117 (2018) 8-20. https://doi.org/10.1016/j.ejps.2018.01.046

[43]   R.A. Rashid, A.H. Jawad, M.A.B.M. Ishak, N.N. Kasim, FeCl3 -activated carbon developed from coconut leaves: Characterization and application for methylene blue removal, Sains Malaysiana 47 (2018) 603-610. https://doi.org/10.17576/jsm-2018-4703-22

[44]   N.S.A. Shukor, A.B. Alias, M.A.M. Ishak, R.R.R. Deris, A.H. Jawad, K.A. Radzun, K. Ismail, Sulfur dioxide gas adsorption study using mixed activated carbon from different biomass, International Journal of Technology 9 (2018) 1121-1131. https://doi.org/10.14716/ijtech.v9i6.2358

[45]   X. Xiong, I.K.M. Yu, L. Cao, D.C.W. Tsang, S. Zhang, Y.S. Ok, A review of biochar-based catalysts for chemical synthesis, biofuel production, and pollution control, Bioresource Technology 246 (2017) 254-270. https://doi.org/10.1016/j.biortech.2017.06.163

[46]   K.-W. Jung, K.-H. Ahn, Fabrication of porosity-enhanced MgO/biochar for removal of phosphate from aqueous solution: Application of a novel combined electrochemical modification method, Bioresource Technology 200 (2016)1029-1032. https://doi.org/10.1016/j.biortech.2015.10.008

[47]   X. Dong, L. He, H. Hu, N. Liu, S. Gao, Y. Piao, Removal of 17β-estradiol by using highly adsorptive magnetic biochar nanoparticles from aqueous solution, Chemical Engineering Journal 352 (2018) 371-379. https://doi.org/10.1016/j.cej.2018.07.025

[48]   B. Kavitha, P.V.L. Reddy, B. Kim, S.S. Lee, S.K. Pandey, K.-H. Kim, Benefits and limitations of biochar amendment in agricultural soils: A review, Journal of Environmental Management 227 (2018) 146-154. https://doi.org/10.1016/j.jenvman.2018.08.082

[49]   T. Xu, L. Lou, L. Luo, R. Cao, D. Duan, Y. Chen, Effect of bamboo biochar on pentachlorophenol leachability and bioavailability in agricultural soil, Science of the Total Environment 414 (2012) 727-731. https://doi.org/10.1016/j.scitotenv.2011.11.005

[50]   A. Kumar, A. Kumar, G. Sharma, M. Naushad, R.C. Veses, A.A. Ghfar, F.J. Stadler, M.R. Khan, Solar-driven photodegradation of 17-β-estradiol and ciprofloxacin from waste water and CO2 conversion using sustainable coalchar/ polymeric-g-C3N4/RGO metal-free nano-hybrids, New Journal of Chemistry 41 (2017) 10208-10224. https://doi.org/10.1039/C7NJ01580A

[51]   K. Sun, K. Ro, M. Guo, J. Novak, H. Mashayekhi, B. Xing, Sorption of bisphenol A, 17α-ethinyl estradiol and phenanthrene on thermally and hydrothermally produced biochars, Bioresource technology 102 (2011) 5757-5763. https://doi.org/10.1016/j.biortech.2011.03.038

[52]   R.-k. Xu, S.-c. Xiao, J.-h. Yuan, A.-z. Zhao, Adsorption of methyl violet from aqueous solutions by the biochars derived from crop residues, Bioresource technology 102 (2011) 10293-10298. https://doi.org/10.1016/j.biortech.2011.08.089

[53]   C.-H. Zhou, X. Xia, C.-X. Lin, D.-S. Tong, J. Beltramini, Catalytic conversion of lignocellulosic biomass to fine chemicals and fuels, Chemical Society Reviews 40 (2011) 5588-5617. https://doi.org/10.1039/c1cs15124j

[54]   D.J. Farrelly, C.D. Everard, C.C. Fagan, K.P. McDonnell, Carbon sequestration and the role of biological carbon mitigation: A review, Renewable and Sustainable Energy Reviews 21 (2013) 712-727. https://doi.org/10.1016/j.rser.2012.12.038

[55]   M. Azeem, R. Hayat, Q. Hussain, M. Ahmed, G. Pan, M. Ibrahim Tahir, M. Imran, M. Irfan, H. Mehmood ul, Biochar improves soil quality and N2-fixation and reduces net ecosystem CO2 exchange in a dryland legume-cereal cropping system, Soil and Tillage Research 186 (2019) 172-182. https://doi.org/10.1016/j.still.2018.10.007

[56]   W.-J. Liu, H. Jiang, H.-Q. Yu, Development of Biochar-Based Functional Materials: Toward a Sustainable Platform Carbon Material, Chemical Reviews 115 (2015) 12251-12285. https://doi.org/10.1021/acs.chemrev.5b00195

[57]   H. Fu, S. Ma, P. Zhao, S. Xu, S. Zhan, Activation of peroxymonosulfate by graphitized hierarchical porous biochar and MnFe2O4 magnetic nanoarchitecture for organic pollutants degradation: Structure dependence and mechanism, Chemical Engineering Journal 360 (2019) 157-170. https://doi.org/10.1016/j.cej.2018.11.207

[58]   S. Wang, Y. Zhou, S. Han, N. Wang, W. Yin, X. Yin, B. Gao, X. Wang, J. Wang, Carboxymethyl cellulose stabilized ZnO/biochar nanocomposites: Enhanced adsorption and inhibited photocatalytic degradation of methylene blue, Chemosphere 197 (2018) 20-25. https://doi.org/10.1016/j.chemosphere.2018.01.022

[59]   M. Chen, C. Bao, D. Hu, X. Jin, Q. Huang, Facile and low-cost fabrication of ZnO/biochar nanocomposites from jute fibers for efficient and stable photodegradation of methylene blue dye, Journal of Analytical and Applied Pyrolysis 139 (2019) 319-332. https://doi.org/10.1016/j.jaap.2019.03.009

[60]   J. Nelson, E.G. Griffin, ADSORPTION OF INVERTASE, Journal of the American Chemical Society 38 (1916)1109-1115. https://doi.org/10.1021/ja02262a018

[61]   J. Schückel, A. Matura, K.-H. Van Pee, One-copper laccase-related enzyme from Marasmius sp.: Purification, characterization and bleaching of textile dyes, Enzyme and

microbial technology 48 (2011) 278-284.
https://doi.org/10.1016/j.enzmictec.2010.12.002

[62]   B. Ismail, S. Nielsen, Invited review: plasmin protease in milk: current knowledge and relevance to dairy industry, Journal of dairy science 93 (2010) 4999-5009. https://doi.org/10.3168/jds.2010-3122

[63]   Y. Bai, H. Huang, K. Meng, P. Shi, P. Yang, H. Luo, C. Luo, Y. Feng, W. Zhang, B. Yao, Identification of an acidic α-amylase from Alicyclobacillus sp. A4 and assessment of its application in the starch industry, Food Chemistry 131 (2012) 1473-1478. https://doi.org/10.1016/j.foodchem.2011.10.036

[64]   T.K. Hakala, T. Liitiä, A. Suurnäkki, Enzyme-aided alkaline extraction of oligosaccharides and polymeric xylan from hardwood kraft pulp, Carbohydrate polymers 93 (2013) 102-108. https://doi.org/10.1016/j.carbpol.2012.05.013

[65]   C.S. Rao, T. Sathish, P. Ravichandra, R. Prakasham, Characterization of thermo- and detergent stable serine protease from isolated Bacillus circulans and evaluation of eco-friendly applications, Process Biochemistry 44 (2009) 262-268. https://doi.org/10.1016/j.procbio.2008.10.022

[66]   K. Luo, Q. Yang, J. Yu, X.-m. Li, G.-j. Yang, B.-x. Xie, F. Yang, W. Zheng, G.-m. Zeng, Combined effect of sodium dodecyl sulfate and enzyme on waste activated sludge hydrolysis and acidification, Bioresource technology 102 (2011) 7103-7110. https://doi.org/10.1016/j.biortech.2011.04.023

[67]   Z. Tong, Z. Qingxiang, H. Hui, L. Qin, Z. Yi, Removal of toxic phenol and 4-chlorophenol from waste water by horseradish peroxidase, Chemosphere 34 (1997) 893-903. https://doi.org/10.1016/S0045-6535(97)00015-5

[68]   L. Cao, Immobilised enzymes: science or art?, Current Opinion in Chemical Biology 9 (2005) 217-226. https://doi.org/10.1016/j.cbpa.2005.02.014

[69]   D.T. Mitchell, S.B. Lee, L. Trofin, N. Li, T.K. Nevanen, H. Söderlund, C.R. Martin, Smart nanotubes for bioseparations and biocatalysis, Journal of the American Chemical Society 124 (2002) 11864-11865. https://doi.org/10.1021/ja027247b

[70]   J.M. Thomas, J. Yordy, J. Amador, M. Alexander, Rates of dissolution and biodegradation of water-insoluble organic compounds, Applied and Environmental Microbiology 52 (1986) 290-296. https://doi.org/10.1128/aem.52.2.290-296.1986

[71]   F.C. Fraga, A. ValÃ©rio, V.A. de Oliveira, M. Di Luccio, D.b. de Oliveira, Effect of magnetic field on the EversaÂ® Transform 2.0 enzyme: Enzymatic activity and structural conformation, International journal of biological macromolecules 122 653-658. https://doi.org/10.1016/j.ijbiomac.2018.10.171

[72]   M.A. Rao, R. Scelza, F. Acevedo, M.C. Diez, L. Gianfreda, Enzymes as useful tools for environmental purposes, Chemosphere 107 (2014) 145-162. https://doi.org/10.1016/j.chemosphere.2013.12.059

[73]   P. Calza, P. Avetta, G. Rubulotta, M. Sangermano, E. Laurenti, TiO2-soybean peroxidase composite materials as a new photocatalytic system, Chemical Engineering Journal 239 (2014) 87-92. https://doi.org/10.1016/j.cej.2013.10.098

[74]   X. Chen, W. Lu, T. Xu, N. Li, D. Qin, Z. Zhu, G. Wang, W. Chen, A bio-inspired strategy to enhance the photocatalytic performance of g-C3N4 under solar irradiation by axial coordination with hemin, Applied Catalysis B: Environmental 201 (2017) 518-526. https://doi.org/10.1016/j.apcatb.2016.08.020

[75]   M.V. Arularasu, J. Devakumar, T.V. Rajendran, An innovative approach for green synthesis of iron oxide nanoparticles: Characterization and its photocatalytic activity, Polyhedron 156 (2018) 279-290. https://doi.org/10.1016/j.poly.2018.09.036

[76]   Huang J. H, Ho W. K, Wang X. C, Metal-free disinfection effects induced by graphitic carbon nitride polymers under visible light illumination. Chem Commun., 50 (2014) 4338–4340. https://doi.org/10.1039/c3cc48374f

[77]   A. Angel Ezhilarasi, J. Judith Vijaya, L. John Kennedy, K. Kaviyarasu, Green mediated NiO nano-rods using Phoenix dactylifera (Dates) extract for biomedical and environmental applications, Mater. Chem. Phys., 241 (2020) 122419. https://doi.org/10.1016/j.matchemphys.2019.122419

[78]   G. Fu, P.S. Vary, C.T. Lin, Anatase TiO_2 nanocomposites for antimicrobial coatings, J. Phys. Chem. B 109 (2005) 8889–8898. https://doi.org/10.1021/jp0502196

[79]   A. P. L. Kvitek, R. Prucek, M. Kolar, R. Vecerova, N. Pizurova, V. K. Sharma, T. Nevecna, R. Zboril, J. Phys. Chem. B 110 (2006) 16248–16253. https://doi.org/10.1021/jp063826h

[80]   M. M. K. Motlagh, A. A. Youzbashi, L. Sabaghzadeh, Synthesis and characterization of Nickel hydroxide/oxide nanorods by the complexation-precipitation method, Int. J. Phys. Sci. 6 (2011) 1471–1476.

[81]   J. R. Morones, J. L. Elechigierra, A. Caacho, K. Holt, J. B. Kouri, J. T. Ramirez, M. J. Yacaman, The bactericidal effect of silver nanorods, J.Nanotechnol. 16 (2005) 2346–2353. https://doi.org/10.1088/0957-4484/16/10/059

[82]   M. Cho, H. Chung, W. Choi, J. Yoon, LinearCorrelationbetween inactivation of E. coli and OH radical concentration in TiO_2 Photocatalytic disinfection, Water Res. 38 (2004) 1069–1077. https://doi.org/10.1016/j.watres.2003.10.029

[83]   J. Sawai, E. Kawada, F. Kanou, H. Igarashi, A. Hashimoto, T. Kokugan, M. Shimizu, Detection of active Oxygen Generated from ceramic powders having Antibacterial Activity, J. Chem. Eng. Jpn. 29 (1996) 627–633. https://doi.org/10.1252/jcej.29.627

[84]   J. Du, J. M. Gebicki, Proteins are major initial cell targets of Hydroxyl free radicals, Int. J. Biochem. Cell Biol. 36 (2004) 2334–2343. https://doi.org/10.1016/j.biocel.2004.05.012

[85]   Seerangaraj Vasantharaj, Selvam Sathiyavimal, Palanisamy Senthilkumar, Felix LewisOscar, Arivalagan Pugazhendhi, Biosynthesis of iron oxide nanoparticles using leaf extract of Ruellia tuberose: Antimicrobial properties and their applications in photocatalytic Degradation, J. Photoch. Photobio. B, 192 (2019) 74–82. https://doi.org/10.1016/j.jphotobiol.2018.12.025

[86]   S. C. Sharma, ZnO nano-flowers from *Carica papaya* milk: Degradation of Alizarin Red-S dye and antibacterial activity against *Pseudomonas aeruginosa* and *Staphylococcus aureus* , Optik, 127 (2016) 6498-6512. https://doi.org/10.1016/j.ijleo.2016.04.036

[87]   R. Brayner, R. Ferrari Iliou, N. Brivois, S. Djediat, M.F. Benedetti, F. Fievet Toxicological Impact Studies Based on Escherichia coli Bacteria in Ultrafine ZnO Nanoparticles Colloidal Medium, Nano Lett. 6 (2006) 866-870. https://doi.org/10.1021/nl052326h

[88]   G. Fu, P. S. Vary, C. T. Lin, Anatase TiO2 Nanocomposites for Antimicrobial Coatings, J. Phys. Chem. B 109 (2005) 8889-8898. https://doi.org/10.1021/jp0502196

[89]   Fahim Hossain, Oscar J. Perales-Perez, Sangchul Hwang, Félix Román, Antimicrobial nanomaterials as water disinfectant: Applications, limitations and future perspectives, Sci. Total Environ., 466–467 (2014) 1047–1059. https://doi.org/10.1016/j.scitotenv.2013.08.009

[90]   Markowska-Szczupak. A, Ulfig. K, Morawski. A. W, The application of titanium dioxide for deactivation of bioparticulates: an overview, Catal. Today, 169 (2011) 257–69. https://doi.org/10.1016/j.cattod.2010.11.055

[91]   Li Q, Mahendra S, Lyon DY, Brunet L, Liga MV, Li D, et al., Antimicrobial nanomaterials for water disinfection and microbial control: potential applications and implications, Water Res., 42(18) (2008) 4591–602. https://doi.org/10.1016/j.watres.2008.08.015

[92]   Wei C, Lin WY, Zainal Z, Williams NE, Zhu K, Kruzic AP, et al., Bactericidal activity of TiO$_2$ photocatalyst in aqueous media: toward a solar assisted water

disinfection system, Environ. Sci. Technol. 28 (1994) 934–8.
https://doi.org/10.1021/es00054a027

[93]   Sun D. D, Tay J. H, Tan K. M, Photocatalytic degradation of E. coli form in water, Water Res., 37 (2003) 3452–62. https://doi.org/10.1016/S0043-1354(03)00228-8

[94]   Cho M, Chung H, ChoiW, Yoon J., Different inactivation behavior of MS-2 phase and E. coli in TiO2 photocatalytic disinfection, Appl. Environ. Microbiol., 71(1) (2005) 270–5. https://doi.org/10.1128/AEM.71.1.270-275.2005

[95]   Choi H, Antoniou M. G, De la Cruz A. A, Stathatos E, Dionysiou D. D, Photocatalytic TiO$_2$ films and membranes for the development of efficient wastewater treatment and reuse systems, Desalination, 202 (2006) 199–206.
https://doi.org/10.1016/j.desal.2005.12.055

[96]   Liu Y, Li J, Qiu X, Burda C, Bactericidal activity of nitrogen-doped metal oxide nanocrystals and the influence of bacterial extracellular polymeric substances (EPS), J. Photochem. Photobiol. A - Chem. 190 (2007) 94–100.
https://doi.org/10.1016/j.jphotochem.2007.03.017

[97]   Deckers A. S, Loo S, L'Hermite M. M, Boime N. H, Menguy N, Reynaud C, et al., Size-, composition- and shape-dependent toxicological impact of metal oxide nanoparticles and carbon nanotubes towards bacteria, Environ. Sci. Technol. 43 (2009) 8423–9. https://doi.org/10.1021/es9016975

[98]   Liga M. V, Bryant E. L, Colvin V. L, Li Q, Virus inactivation by silver doped titanium dioxide nanoparticles for drinking water treatment, Water Res., 45 (2011) 535–44. https://doi.org/10.1016/j.watres.2010.09.012

[99]   Dimitroula H, Daskalaki V. M, Frontistis Z, Kondarides D, Panagiotopoulou P, Xekoukoulotakis N. P, et al., Solar photocatalysis for the abatement of emerging micro contaminants in wastewater: synthesis, characterization and testing of various TiO2 samples, Appl. Catal. Environ., 117–118 (2012) 283–91.
https://doi.org/10.1016/j.apcatb.2012.01.024

[100] El-Sayed R. El-Sayed, Heba K. Abdelhakim, Zainab Zakaria, Extracellular biosynthesis of cobalt ferrite nanoparticles by *Monascus purpureus*and their antioxidant, anticancer and antimicrobial activities: Yield enhancement by gamma irradiation, Mater. Sci. Eng. C. Mater. Biol. Appl., 107 (2020) 110318.
https://doi.org/10.1016/j.msec.2019.110318

[101] Yi Li, Chi Zhang, Danmeng Shuai, Saraschandra Naraginti, Dawei Wang, Wenlong Zhang, Visible-light-driven photocatalytic inactivation of MS2 by metal-free g-C3N4: Virucidal performance and mechanism, Water Research 106 (2016) 249-258.
https://doi.org/10.1016/j.watres.2016.10.009

[102] Tura Safawo, B. V Sandeep, Sudhakar Pola, Aschalew Tadesse, Synthesis and characterization of zinc oxide nanoparticles using tuber extract of anchote (Coccinia abyssinica (Lam.) Cong.) for antimicrobial and antioxidant activity assessment, OpenNano, 3 (2018) 56–63. https://doi.org/10.1016/j.onano.2018.08.001

[103] P. J. P. Espitia, N. deF. F. Soares, J. S. dos R. Coimbra, N. J. de Andrade, R. S. Cruz, E. A. A. Medeiros, Zinc oxide nanoparticles: synthesis, antimicrobial activity and food packaging applications, Food Bioprocess Technol 5 (2012) 1447–1464. https://doi.org/10.1007/s11947-012-0797-6

[104] M. Premanathan, K. Karthikeyan, K. Jeyasubramanian, G. Manivannan, Selective toxicity of ZnO nanoparticles toward Gram-positive bacteria and cancer cells by apoptosis through lipid peroxidation, nanomedicine nanotechnology, Biol. Med. 7 (2011) 184–192. https://doi.org/10.1016/j.nano.2010.10.001

[105] K. R. Raghupathi, R. T. Koodali, A. C. Manna, Size-dependent bacterial growth inhibition and mechanism of antibacterial activity of zinc oxide nanoparticles, Langmuir 27 (2011) 4020–4028. https://doi.org/10.1021/la104825u

[106] C. Jayaseelan, A.A. Rahuman, G. Rajakumar, A. Vishnu Kirthi, T. Santhoshkumar, S. Marimuthu, A. Bagavan, C. Kamaraj, A.A. Zahir, G. Elango, Synthesis of pediculocidal and larvicidal silver nanoparticles by leaf extract from heartleaf moonseed plant, Tinospora cordifolia Miers, Parasitol. Res. 109 (2011) 185–194. https://doi.org/10.1007/s00436-010-2242-y

[107] R. Venckatesh, P. Rajiv, S. Rajeshwari, Bio-fabrication of zinc oxide nanoparticles using leaf extract of Parthenium hysterophorus L. and its size-dependent antifungal activity against plant fungal pathogens. Spectrochim. Acta A Mol. Biomol. Spectrosc. 112 (2013) 384–387. https://doi.org/10.1016/j.saa.2013.04.072

[108] N. Jones, B. Ray, K. T. Ranjit, A. C. Manna, Antibacterial activity of ZnO nanoparticle suspensions on a broad spectrum of microorganisms, FEMS Microbiol. Lett. 279 (2008) 71–76. https://doi.org/10.1111/j.1574-6968.2007.01012.x

[109] R. Brayner, R. Ferrari-Iliou, N. Brivois, S. Djediat, M.F. Benedetti, F. Fiévet, Toxicological impact studies based on Escherichia coli bacteria in ultrafine ZnO nanoparticles colloidal medium, Nano Lett. 6 (2006) 866–870. https://doi.org/10.1021/nl052326h

[110] Xiuquan Xua, Songmei Wang, Xiaofeng Yua, Jila Dawa, Dongliang Gui, Ronghui Tang, Biosynthesis of Ag deposited phosphorus and sulfur co-doped g-C$_3$N$_4$ with enhanced photocatalytic inactivation performance under visible light, Appl. Surf. Sci., 501 (2020) 144245. https://doi.org/10.1016/j.apsusc.2019.144245

[111] T. Feng, J. L. Liang, Z. Y. Ma, M. Li, M. P. Tong, Bactericidal activity and mechanisms of BiOBr-AgBr under both dark and visible light irradiation conditions, Colloid. Surface B 167 (2018) 275–283. https://doi.org/10.1016/j.colsurfb.2018.04.022

[112] B. Pant, P. Pokharel, A. P. Tiwari, P. S. Saud, M. Park, Z. K. Ghouri, S. Choi, S. J. Park, H. Y. Kim, Characterization and antibacterial properties of aminophenol grafted and Ag NPs decorated graphene nanocomposites, Ceram. Int. 41 (2015) 5656–5662. https://doi.org/10.1016/j.ceramint.2014.12.150

[113] W. Bing, Z. W. Chen, H. J. Sun, P. Shi, N. Gao, J. S. Ren, X. G. Qu, Visible-light-driven enhanced antibacterial and biofilm elimination activity of graphitic carbon nitride by embedded Ag nanoparticles, Nano Res. 8 (2015) 1648–1658. https://doi.org/10.1007/s12274-014-0654-1

[114] M. Arunpandian, K. Selvakumar, A. Raja, M. Thiruppathi, P. Rajasekaran, P. Rameshkumar, E. R. Nagarajan, S. Arunachalam, Development of novel Nd_2WO_6/ ZnO incorporated on GO nanocomposite for the photocatalytic degradation of organic pollutants and biological studies, J. Mater. Sci.: Mater. Electron. 30 (2019) 18557-18574. https://doi.org/10.1007/s10854-019-02209-9

[115] Y. S. Fu, H. Ting, L. L. Zhang, J. W. Zhu, X. Wang, Ag/g-C_3N_4 catalyst with superior catalytic performance for the degradation of dyes: a borohydride-generated superoxide radical approach, Nanoscale 7 (2015) 13723. https://doi.org/10.1039/C5NR03260A

[116] P. K. Stoimenov, R. L. Klinger, G. L. Marchin, K. J. Klabunde, Metal oxide nanoparticles as bactericidal agents, Langmuir 18 (2002) 6679–6686. https://doi.org/10.1021/la0202374

[117] T. Hamouda, J. R. Baker, Antimicrobial mechanism of action of surfactant lipid preparations in centeric Gram-negative bacilli, J. Appl. Microbiol. 89 (2000) 397–403. https://doi.org/10.1046/j.1365-2672.2000.01127.x

[118] I. Sondi, B. S. Sondi, Silver nanoparticles as antimicrobial agent: a case study on E. coli as a model for Gram-negative bacteria, J. Colloid Interface Sci. 275 (2004) 177–182. https://doi.org/10.1016/j.jcis.2004.02.012

[119] S. S. Lee,W. Song, M. Cho, H. L. Puppala, P. Nguyen, H. Zhu, L. Segatori, V. L. Colvin, Antioxidant properties of cerium oxide nanocrystals as function of nanocrystals diameterand surface coating, ACS Nano. 7 (2013) 9693-9703. https://doi.org/10.1021/nn4026806

[120] Assumpta Chinwe Nwanya, Lovasoa Christine Razanamahandry, A. K. H. Bashir, Chinwe O. Ikpo, Stephen C. Nwanya, Subelia Botha, S. K. O. Ntwampe, Fabian I. Ezema, Emmanuel I. Iwuoha, Malik Maaza, Industrial textile effluent treatment and

antibacterial effectiveness of Zea mays L. Dry husk mediated bio-synthesized copper oxide nanoparticles, J. Hazard., 375 (2019) 281–289. https://doi.org/10.1016/j.jhazmat.2019.05.004

[121] Qu, Y. Q. & Duan, X. F. Progress, challenge and perspective of heterogeneous photocatalysts. Chem. Soc. Rev.42 (2013)2568–2580. https://doi.org/10.1039/C2CS35355E

[122] Park, H. J., Nguyen, T. T. M., Yoon, J. & Lee, C. Role of reactive oxygen species in *Escherichia coli* inactivation by cupric ion. Environ.Sci. Technol. 46 (2012)11299–11304. https://doi.org/10.1021/es302379q

[123] Wang, W. J. *et al.* Visible-Light-Driven photocatalytic inactivation of *E. coli* K-12 by bismuth vanadate nanotubes: Bactericidal performance and mechanism. Environ. Sci. Technol. 46 (2012)4599–4606. https://doi.org/10.1021/es2042977

[124] Rajesh Pandiyan, Shanmugam Mahalingam, Young-Ho Ahn, Antibacterial and photocatalytic activity of hydrothermally synthesized SnO2 doped GO and CNT under visible light irradiation, J. Photoch. Photobio. B,191 (2019) 18–25. https://doi.org/10.1016/j.jphotobiol.2018.12.007

[125] R. Pandiyan, S. Ayyaru, Y.H. Ahn, Non-toxic properties of TiO2 and STiO2 nanocomposite PES ultrafiltration membranes for application in membrane-based environmental biotechnology, Ecotoxicol. Environ. Saf. 158 (2018) 248–255. https://doi.org/10.1016/j.ecoenv.2018.04.027

[126] M. Naushad, T. Ahamad, G. Sharma, A.A.H. Al-Muhtaseb, A.B. Albadarin, M.M. Alam, Z.A. Alothman, S.M. Alshehri, A.A. Ghfar, Synthesis and characterization of a new starch/SnO2 nanocomposite for efficient adsorption of toxic Hg^{2+} metal ion, Chem. Eng. J. 300 (2016) 306–316. https://doi.org/10.1016/j.cej.2016.04.084

[127] A. Fakhri, S. Behrouz, M. Pourmand, Synthesis, photocatalytic and antimicrobial properties of SnO2, SnS2 and SnO2/SnS2 nanostructure, J. Photochem. Photobiol. B Biol. 149 (2015) 45–50. https://doi.org/10.1016/j.jphotobiol.2015.05.017

[128] S. Sudhaparimala, M. Vaishnavi, Biological synthesis of nano composite SnO2 efficient photocatalytic degradation and antimicrobial activity, Mater. Today 3 (2016) 2373–2380. https://doi.org/10.1016/j.matpr.2016.04.150

[129] M. A. Qamar, S. Shahid, S. A. Khan, S. Zaman, M. N. Sarwar, Synthesis characterization, optical and antibacterial studies of co-doped SnO2 nanoparticles, Digest J. Nanomater. Biostruct. 12 (2017) 1127–1135.

[130] W. Sangchay, The self-cleaning and photocatalytic properties of TiO_2 doped with SnO_2 thin films preparation by Sol-gel Method, Energy Procedia 89 (2016) 170–176. https://doi.org/10.1016/j.egypro.2016.05.023

[131] S. Kumar, M. Kumar, A. Thakur, S. Patial, Water treatment using photocatalytic and antimicrobial activities of tin oxide nanoparticles, Ind. J. Chem. Tech. 24 (2017) 435–440.

[132] K. A. Omar, B. I. Meena, S. A. Muhammed, Study on the activity of ZnO-SnO_2 nanocomposite against bacteria and fungi, Physicochem. Probl. Miner. Process 52 (2016) 754–766.

[133] K. Kaviyarasu, C. Maria Magdalane, K. Kanimozhi, J. Kennedy, B. Siddhardha, E. Subba Reddy, Naresh Kumar Rotte, Chandra Shekhar Sharma, F.T. Thema, Douglas Letsholathebe, Genene Tessema Mola, M. Maaza, Elucidation of photocatalysis, photoluminescence and antibacterial studies of ZnO thin films by spin coating method, J. Photochem. Photobiol. B, 173 (2017) 466-475. https://doi.org/10.1016/j.jphotobiol.2017.06.026

[134] K. Karthik, S. Dhanuskodi, C. Gobinath, S. Prabukumar, S. Sivaramakrishnan, Fabrication of MgO nanostructures and its efficient photocatalytic, antibacterial and anticancer performance, J. Photochem. Photobiol. B, 190 (2019) 8-20. https://doi.org/10.1016/j.jphotobiol.2018.11.001

[135] K. Karthik, S. Dhanuskodi, C. Gobinath, S. Prabukumar, S. Sivaramakrishnan, *Andrographis paniculata* extract mediated green synthesis of CdO nanoparticles and its electrochemical and antibacterial studies, J. Mater. Sci: Mater. Electron. 28 (2017) 7991-8001. https://doi.org/10.1007/s10854-017-6503-8

[136] K. Karthik, S. Dhanuskodi, C. Gobinath, S. Sivaramakrishnan, Microwave-assisted synthesis of CdO-ZnO nanocomposite and its antibacterial activity against human pathogens, Spectrochimic. Acta Part A: Mol. Biomol. Spectrosc. 139 (2015) 7-12. https://doi.org/10.1016/j.saa.2014.11.079

Bioinspired Nanomaterials for Energy and Environmental Applications   Materials Research Forum LLC
Materials Research Foundations **121** (2022) 83-116   https://doi.org/10.21741/9781644901830-3

Chapter 3

Bioinspired Nanomaterials for Photocatalytic Degradation of Toxic Chemicals

Vellaichamy Balakumar[1,*], Ramalingam Manivannan[2] and Keiko Sasaki[1]

[1]Department of Earth Resources Engineering, Faculty of Engineering, Kyushu University, 744 Motooka, Nishiku, Fukuoka 819-0395, Japan

[2]Department of Advanced Organic Materials Engineering, Chungnam National University, 220 Gung-dong, Yuseong-gu, Daejeon, 305-764, South Korea

*chembalakumar@gmail.com (V. Balakumar)

Abstract

This chapter focuses on photocatalysts synthesized based on plant extracts (stems, leaves, flowers, roots), microorganisms, and natural oils. This chapter presents various bio-inspired nanomaterials like metal, metal oxide nanoparticles, metal and metal oxide-based nanocomposite, with different morphologies, modifications, and recent progress advances. The inspired nanomaterials, which can be used as sustainable photocatalysts for degrading toxic chemicals and deliberately discussed further highlight this catalyst's enormous potential. The role of catalysts, which includes surface area, charge separation, and biocompatibility, have also been elaborately explained. Moreover, future challenges and opportunities are included in this chapter. The information is expected to be comprehensive and valuable to the chemistry and materials science community interested in bioinspired photocatalysts for environmental remediation and beyond.

Keywords

Bioinspired Nanomaterials, Synthesis, Photocatalytic Degradation, Charge Separation Mechanism, Toxic Chemicals

Contents

Bioinspired Nanomaterials for Photocatalytic Degradation of Toxic Chemicals **83**

1. **Introduction**...**84**

2.   General principles and charge separation of
semiconductor photocatalysis 86

3.   Synthesis of plant-mediated nanomaterials 86

4.   Bio-mediated noble metal nanoparticles and its photocatalytic
degradation towards toxic chemicals 88

5.   Bio-mediated transition metal nanoparticles and its
photocatalytic degradation towards toxic chemicals...90

6.   Bio-mediated metal nanoparticles based nanocomposites for
photocatalytic degradation of toxic chemicals 91

7.   Bio-mediated ZnO nanoparticles and its photocatalytic
degradation towards toxic chemicals 92

8.   Bio-mediated iron oxide nanoparticles and its photocatalytic
degradation towards toxic chemicals 94

9.   Bio-mediated other metal oxide nanoparticles and its
photocatalytic degradation towards toxic chemicals...96

10.   Bio-mediated metal oxide based nanocomposites for
photocatalytic degradation of toxic chemicals 99

11.   Microorganism mediated nanomaterials for photocatalytic
degradation of toxic chemicals..100

Conclusion and future outlook 101

Acknowledgments 102

References 102

1.   Introduction

In the survival of any form of life, water is an essential constituent, and the earth surrounds 70% of water. However, all forms of public water consumption are only 2.5% [1]. In the 21st century, environmental remediation is the highest challenge for researchers due to industrial growth, uncontrolled groundwater development, global warming, and climate changes [2, 3, 4]. Organic dyes and other toxic pollutants are discharged from various industrial sectors [5, 6, 7, 8]. Worldwide, 106 tons of synthetic dyes are produced and released into the environment as wastewaters. It is well known that even a minimum concentration of dye pollutants can create problems in aquatic environments. It is reported that highly toxic to human beings and animals. Therefore, removing these pollutants and protecting the environment are substantial [5]. Until date, various technologies including adsorption [9], coagulation [10], chemical oxidation [11], membrane separation [12], flotation [13], ion exchange [14], reverse osmosis [15], electrochemical treatment [16], catalytic conversion [17, 18, 19, 20, 21] and photocatalytic degradation [6, 7, 8, 22, 23] have been employed in wastewater treatments. Among the available methods, the photocatalysis approach has been considered green and eco-friendly for solar photodegradation. Up to now, researchers have investigated various semiconductor photocatalysis. Potential applications including water reduction and evolution of H_2 and O_2 [24, 25], photoreduction of CO_2 [26, 27], photocatalytic NOx removal [28, 29], organic pollutants degradation [30, 31, 32, 33, 34], and so on.

To date, research has been focused on venomous nanomaterials photocatalysts for energy and environmental applications. Various methodologies synthesize these nanomaterials. Including hydrothermal method [35, 36, 37], solvothermal reaction [38, 39], ultra-sonication method [40, 41], reflux method [42], sol-gel method [43], simple precipitation method [44], and in-situ method [45] have employed to the synthesis of nanomaterials and utilized for various applications primarily in photocatalysis. The synthesis methods use high temperatures, apply radiations, need specialized apparatus, consume energy, and include toxic chemicals hazardous to the environment. It is costly and cumbersome for the large-scale production of nanomaterials for real-time applications. Therefore, researchers have devoted much attention to fabricating nanomaterials through green methods to avoid toxic and hazardous chemicals. The plant-mediated biosynthesis benefits over other biomaterials, as shown in Scheme. 1

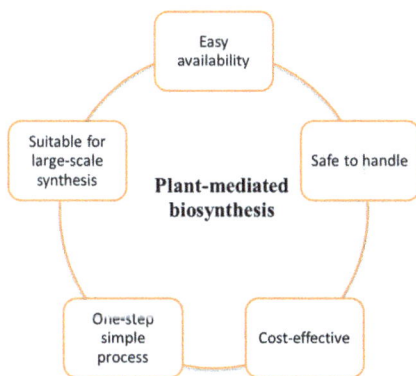

Scheme. 1 Plant-mediated biosynthesis benefits over other biomaterials.

Biological systems (plant extract from leaves, fruits, and seeds and microorganisms like bacteria, fungi, and algae) generally display various biomolecules with reductive properties [46, 47, 48, 49, 50]. This chapter deals with bio-mediated nanomaterials. Concerning their synthesis, the mechanisms involved, and applications in photocatalytic toxic chemicals degradation.

2.   General principles and charge separation of semiconductor photocatalysis

When irradiating with light energy, the photocatalyst initiates or accelerates specific reduction and oxidation processes. In general, three necessary steps involved in a photocatalytic reaction are as follows:

- electron and hole pairs generation

- generated electron and hole pairs separation

- photo-oxidation and photo-reduction reactions take place

Firstly, it is known that the light-harvesting process (step 1) mainly depends on the structure and compact of photocatalysts, which can significantly improve through constructing the meso or microporous. These unique structures are more efficient for utilizing light through its multiple reflections and scattering effects. In the light-harvesting process to facilitate redox reactions. The photogenerated charge carriers can separate holes (h^+) in the valance band (VB). Electrons (e^-) in the photocatalyst surface's conduction band (CB). The photogenerated and separated holes and electrons directly react with O_2 and H_2O on the catalyst's surface to produce superoxide and hydroxyl radicals, respectively. These radicals have substantial reduction and oxidation potentials to degrade aqueous toxic chemicals. All these photocatalytic reaction steps are clearly illustrated in Fig. 1.

3.   Synthesis of plant-mediated nanomaterials

The synthesis of nanomaterials using plant parts such as leaves, fruits, seeds, flowers, roots, stems, and other parts of plants is shown in Fig. 2. Firstly, new plant parts were washed thoroughly, and to remove all the impurities deionized water was used. Then the new features were chopped into small pieces and were boiled (90 °C for 15 min) with deionized water. Finally, the extract was filtered and was used. Another way, plant parts were collected and dried under shade. To contain the section above mentioned method followed. The various phytochemical constituents act as reducing and capping agents. For example, recently, many reports have prepared extracts using multiple plants, including *Crataeva religiosa* [5], *Tagetes erecta* [51], *Azadirachta indica* [52] *Euphorbia hirta* [53]. Finally, the extract solution was used to prepare various nanomaterials.

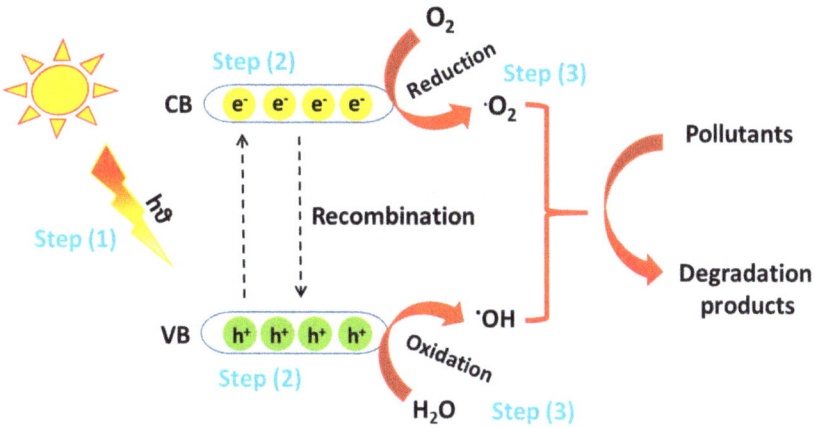

Fig. 1. Illustration of photo activation of semiconductor photocatalysts.

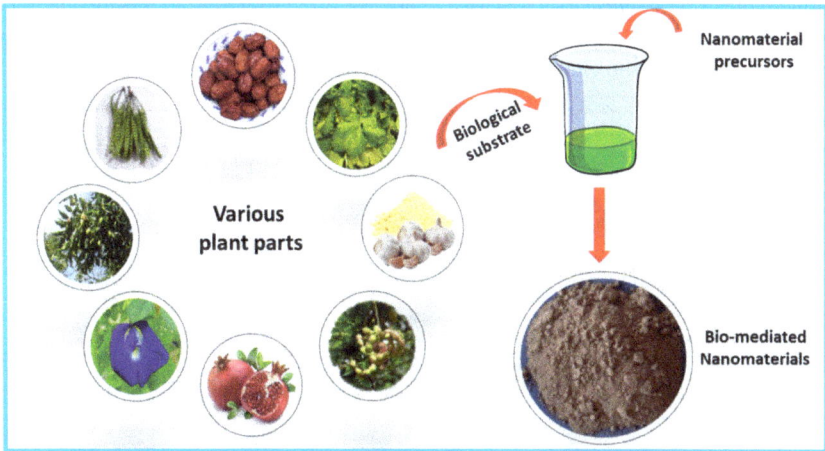

Fig. 2. Green synthesis of nanomaterials using various plant parts.

Bioinspired Nanomaterials for Energy and Environmental Applications   Materials Research Forum LLC
Materials Research Foundations **121** (2022) 83-116   https://doi.org/10.21741/9781644901830-3

4.   Bio-mediated noble metal nanoparticles and its photocatalytic degradation towards toxic chemicals

Noble metal nanoparticles (Pd, Au, Ag) can strongly absorb visible light due to their localized surface plasmon resonance effect and enhance the degradation of dyes. For example, P. Narasaiah et al. [54] investigated the *Pimpinella tirupatiensis* plant extract mediated PdNPs as an excellent photocatalyst for Congo red dye degradation. Further, Pd NPs were investigated using *Catunaregum spinosa (C. Spinosa)* as degradation Congo red with 83% removal efficiency [55]. Recently, A. A. Olajire et al. [56] investigated the biosynthesized PdNPs using *Ananas comosus* leaf extract for polyethylene degradation. Similarly, P. Kumar et al. [57] found AgNPS using *Ulva lactuca* (seaweed) and utilized it for photocatalytic MB dye degradation. Also, cube/square AgNPs (60 nm) were synthesized by phytomethod. The AgNPs show useful photocatalytic decomposition towards MB [58]. Efficient photocatalytic degradation of dyes, MO, Rhodamine B, and Orange G using mature fruits of *Alpinia nigra* mediated silver nanoparticles (AgNPs) [59]. Additionally, gold nanoparticles (AuNPs) were synthesized using *Alpinia nigra* leaves and utilized to degrade MO and Rhodamine B (83.25% and 87.64%) [60]. Another excellent example of green synthesis of AgNPs using *Areca catechu* extract was investigated. This Ag nanocatalyst exhibits acceptable photocatalytic degradation of MB [61]. Furthermore, *Alpinia officinarum* was used to prepare AgNPs through an eco-friendly route [62]. The size range of AgNPs was analyzed and found 50-200 nm. The photodegradation of malachite green and MB dyes was investigated, and the process is shown in Fig. 3.

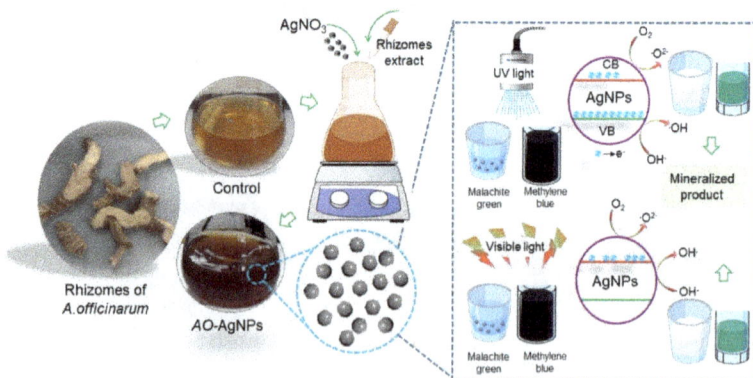

Fig. 3 Biosynthesis AgNPs using Alpinia officinarum rhizome extract and its photocatalytic degradation mechanism. Reproduced permission from Ref. No. [62].

Recently, AuNPs synthesized using *Lagerstroemia speciosa* leaf extract, and it showed strong photocatalytic reduction activity towards organic pollutants under visible light [63]. Biogenic synthesized spherical and triangular AuNPs (10-50 nm) using *Pogestemon benghalensis (B) O. Ktz.* leaf extract used for the MB dye degradation [64]. Ag nanoparticles were synthesized using leaf extract of *Kalanchoe Daigremontiana*, and the Ag particles showed a high photocatalytic activity towards MB (degradation time is one min) as reported by G. A. Molina et al. [65]. Further research is needed to comprehend this phenomenon fully. The detailed comparison is shown in Table 1.

Table 1. *Green synthesis of noble metal nanoparticles and its photocatalytic degradation of toxic chemicals.*

Source	Nanoparticles	Size	Shape/morphology	Photocatalytic degradation	Ref.
Ulva lactuca	AgNPs	48.59 nm	spherical	methyl orange	57
Plukenetia volubilis L.	AgNPs	60 nm	distorted cubic in shape	methylene blue	58
Alpinia nigra fruits	AgNPs	6 nm	spherical	methyl orange, rhodamine B and orange G	59
Alpinia officinarum	AgNPs	50–200 nm	hexagonal	malachite green and methylene blue	62
Alpinia nigra leaves	AuNPs	21.52 nm	spherical	methyl orange and rhodamine B	60
Lagerstroemia speciosa leaf extract	AuNPs	107–193 nm	hexagonal and triangular	methyl orange, methylene blue, and bromocresol green, bromophenol blue and 4-nitrophenol	63
Pogestemon benghalensis leaf extract	AuNPs	10–50 nm	spherical and triangular	methylene blue	64
Actinidia chinensis	FeNPs	91.78–107 nm	cubic and rod	alizarin yellow R	66
Trigonella foenum-graecum seed extract	FeNPs	7–14 nm	spherical	methyl orange	67
Cinnamomum verum bark extract	MnNPs	100 nm	spherical	Congo Red	68
Impatiens balsamina leaf extract	CuNPs	5–10 nm	spherical	methylene blue and congo red	69
Camellia Sinensis	NiNPs	43.87–48.76 nm	spherical	crystal violet	70

5.   Bio-mediated transition metal nanoparticles and its photocatalytic degradation towards toxic chemicals

Transition metal nanoparticles have been considered photocatalysts for environmental remediation due to their excellent reactivity and stability and large surface active sites. For example, A. Ahmed et al. [66] reported the fruit extract of *Actinidia chinensis* (reducing as well as the stabilizing agent) mediated nanosized iron (FeNPs) was done (Fig. 4). The 91.78-107 nm sized FeNPs were cubic and rod-shaped and acted as effective alizarin yellow R dye degradation photocatalyst (Fig. 4). The degradation was achieved 93.7% at 42 h. In addition, a simple green *Trigonella foenum-graecum* seed extract mediated Zero-valent iron (Fe^0) nanoparticles were synthesized, and the size of the nanoparticles was found to be 11 nm. FeNPs are a very effective degradation of MO dye, and the rate constant (k) was found to be 0.025 min^{-1} [67].

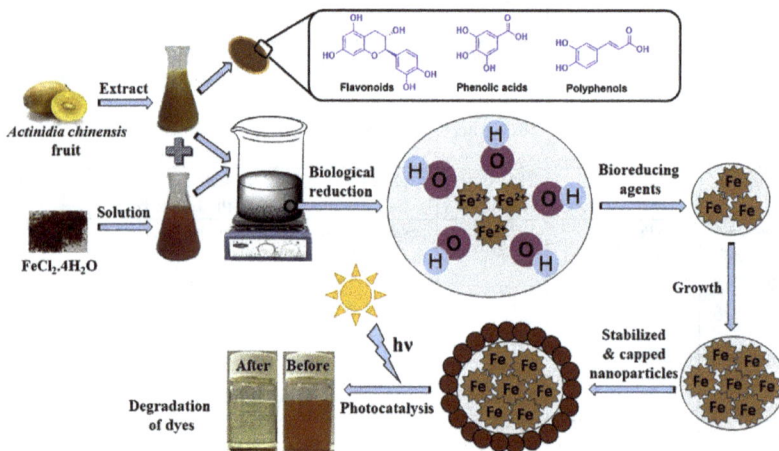

Fig. 4 Bio-mediated FeNPs synthesis using Actinidia chinensis aqueous fruit extract and photocatalytic alizarin yellow R dye degradation. Reproduced permission from Ref. No. [66].

Using manganese(II) acetate tetrahydrate as Mn precursor, U. Kamran et al. [68] fabricated via *Cinnamomum verum* bark extracts mediated manganese nanoparticles (MnNPs). The average size of the synthesized MnNPs was less than 100 nm. At optimized conditions, 78.5% Congo Red dye degradation was achieved using MnNPs. Recently, Roy et al. [69] developed an *Impatiens balsamina* leaf extract as a reducing agent for the synthesis of copper nanoparticles. They were assessed by MB and congo red dye degradation under solar irradiation. Similarly, *Camellia Sinensis* leaves extract was employed to synthesis nickel nanoparticles (NiNPs), and they show excellent photocatalytic activity (99.5%) for the crystal violet dye degradation [70].

6.   Bio-mediated metal nanoparticles based nanocomposites for photocatalytic degradation of toxic chemicals

Two or more nanomaterials composed of hybrid materials are utilized for various applications due to their greater properties over their sole components [17, 18, 19, 20, 21]. For instance, S. Shams et al. [71] developed a novel Au@Fe_2O_3 nanocomposite using an eco-benign method (*Citrus sinensis* fruit extract). In UV-Vis, spectral result nanocomposite showed 290 and 520 nm due to SPR peaks of Fe_2O_3 and Au, respectively. The 94% of MB reduction was achieved within 50 min using Au@Fe_2O_3. Additionally, a reduced graphene oxide(rGO)-AgNPs hybrid nanocomposite was synthesized from an aqueous extract of *Brassica nigra* by simultaneously reducing graphene oxide and silver ions. The hybrid nanocomposite was assessed to remove Direct blue-14 dye photo catalytically [72]. Recently, Panchal et al. [73] reports on investigation of different weight percentage Ag (0.5%, 1.0% and 2.0%) loaded ZnO nanoparticles (Ag/ZnO) using *Ocimum tenuiflorum* (Tulsi) plant seed extract. Further, the synthesized all Ag/ZnO nanocomposite was evaluated for dye degradation. The result showed that 1.0% Ag/ZnO possessed a good photocatalytic activity compared to others. Silver-graphene nanocomposite synthesized with Mortiño (*Vaccinium floribundum Kunth*) berry extract. Showed enhanced photocatalytic activity for the MB, and MO degradation is reported by Karla S. Vizuete et al. [74]. In addition, eco-friendly Ag@CeO2 composites were synthesized and stabilized by garlic extract. The nanocomposites exhibited excellent degradation towards single dyes (MB and MO) and mixed dye (MB + MO). The possible photocatalytic mechanism for the dye's degradation was proposed. The nanocomposite possesses good recyclability and stability [75]. The synthesis scheme and photocatalytic process are illustrated in Fig. 5.

Furthermore, *Ziziphus Jujuba* leaf extract was used to synthesize gold-ZnO nanorods (Au@ZnO). The Au@ZnO exhibited significant photocatalytic activity compared to Au and ZnO NRs towards the photodegradation of MO. Due to the solid visible light absorbance and effective Schottky barrier between the AuNPs and ZnO NRs. Finally, the generated hydroxide and superoxide radicals were investigated through a scavenger study [76].

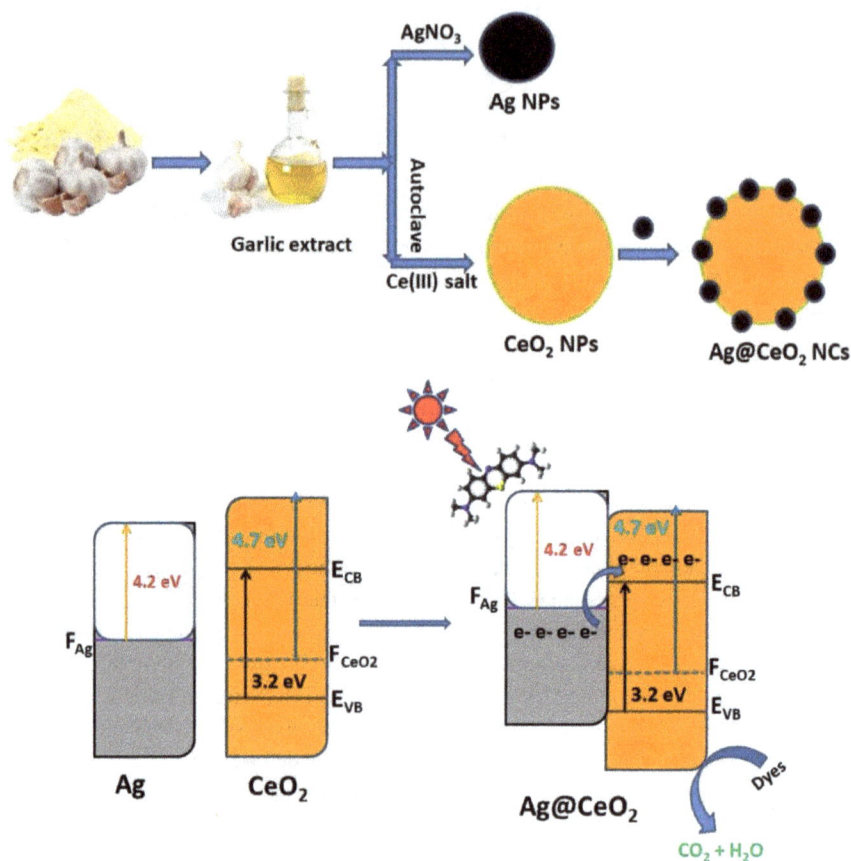

Fig. 5 (A) Preparation of Ag@CeO₂ composites using garlic extract and (B) The photocatalytic mechanism for the degradation. Reproduced with permission from Ref. [75].

7. Bio-mediated ZnO nanoparticles and its photocatalytic degradation towards toxic chemicals

TiO_2 has been the most studied photocatalyst for the degradation of all pollutants. However, TiO_2 has certain drawbacks for the photocatalytic community. Based on this, researchers have shifted towards ZnO as photocatalyst due to the similar bandgap energy (3.2 eV) of TiO_2 and expected to be similar activity. ZnO is an environmentally friendly material and comparatively cheaper than TiO^2. It has good compatibility with a wide range of daily life applications [77]. Compared to conventional methods (hydrothermal, microwave, precipitation, microemulsion, solvothermal, and mechanical techniques). The bio-mediated synthesis of ZnO nanoparticles is based on more availability, cost-effective and straightforward methodology. In the last years, plant extracts that mediated the biosynthesis of various NPs have been an excellent research topic for scientists. M. Golmohammadi et al. [78] developed ZnO nanoparticles using *jujube* fruit extract. They examined the direct sunlight photocatalytic degradation of methylene blue and eriochrome black-T. The degradation efficiencies were observed at 92% and 86% within 5 h and the kinetic constants of 8.7 x 10-3 min-1 and 6.7 x 10-3 min-1 for MB and ECBT, respectively. Recently, ZnO nanoparticles using fruit (*Prunus*) extract were reported by U. Singhal et al. [79]. The photocatalytic degradation of a hazardous dye, MB, was confirmed by the defendant's UV-Vis analysis. The average of 25 nm ZnO nanoparticles. With relatively uniform crystals were prepared using *Quince seed Mucilage* and studied photocatalytic MB degradation [80]. ZnO-NPs with a size range of 25-90 nm were prepared using *Trianthema portulacastrum* extract. Evaluated the photocatalytic activities towards water-soluble synozol navy blue K-BF [81]. Additionally, ZnO nanoparticles were synthesized using *Pithecellobium dulce* peel extract. The ZnO exhibited excellent photodegradation efficiency towards MB and a kinetic rate constant of 0.00812 min^{-1} [82]. The synthesis scheme and photocatalytic mechanism as shown in Fig. 6.

Furthermore, ZnO was prepared using *Corymbia citriodora* leaf extract reported by Zheng et al. [83] and investigated the degradation of MB. The *corriandrum sativum* leaf extract-mediated ZnO nanoparticles were prepared. The as-prepared ZnO shows excellent photocatalytic degradation of anthracene, about 96 % [84]. *Calliandra haematocephala* leaf extract mediated ZnO nanoparticles and obtained an estimated specific surface area of 9.18 m2/g. Under solar radiation, MB degradation was estimated using ZnO [85]. L. Chen et al. [86] reported A *Scutellaria baicalensis (S. baicalensis)* root extract as a reducing agent for synthesizing ZnO and its photocatalytic degradation activity using MB. A *Cassia fistula* plant extract-based ZnO nanoparticles and the formation of nanoparticle sizes were found to be ~5-15 nm. Finally, the ZnO was assessed for effectively degraded MB dye under UV and Sunlight [87]. Similarly, J. Lu et al. [88] reported a one-pot green synthesis of ZnO nanoparticles using *Codonopsis lanceolata* roots. The ZnO NPs were applied to degrade MB dye under UV light (365 nm), and 90.3% degradation was achieved within 40 min. The synthesis of spherical ZnO NPs by *Eucalyptus globulus* leaf extract is also utilized to degrade MB and MO [89]. B. Praveenkumar et al. [90] report on a simple and green ethanolic extract of *typha latifolia.L* leaf mediated zinc oxide nanoparticles. It showed better photo-degradation efficiency. In addition, ZnO nanoparticles were successfully prepared by *Allium sativum* (garlic), *Allium cepa* (onion), and *Petroselinum crispum* (parsley) [91], Kalopanax septemlobus [92], Suaeda japonica extract [93], and Azadirachta indica (Neem) leaf extract [94]. They were used for photocatalytic removal of various toxic chemicals.

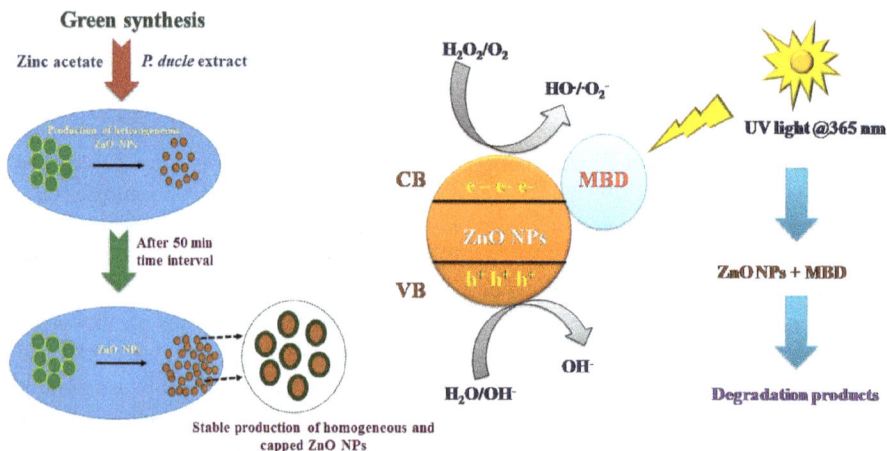

Fig. 6 Green synthesis of ZnO nanoparticles and its photocatalytic mechanism for degradation of methylene blue. Reproduced with permission from Ref. [82].

8.   Bio-mediated iron oxide nanoparticles and its photocatalytic degradation towards toxic chemicals

Recently, biological synthesis of iron oxide nanoparticles by *Lagenaria siceraria* [95], *Camellia sinensis* [96], *Glycosmi mauritiana* [97], and *Eichhornia crassipes* [98] plant extract. In practical application, it is revealed that the nanoparticles are different in properties synthesized by chemical method—especially photodegradation properties compared with biologically synthesized nanoparticles. For instance, S. Qasim et al. [99] investigated the degradation of safranin dye using *Withania coagulans* extract-based iron oxides showed better performances than chemically synthesized iron oxides nanorods, as shown in Fig.7a&b.

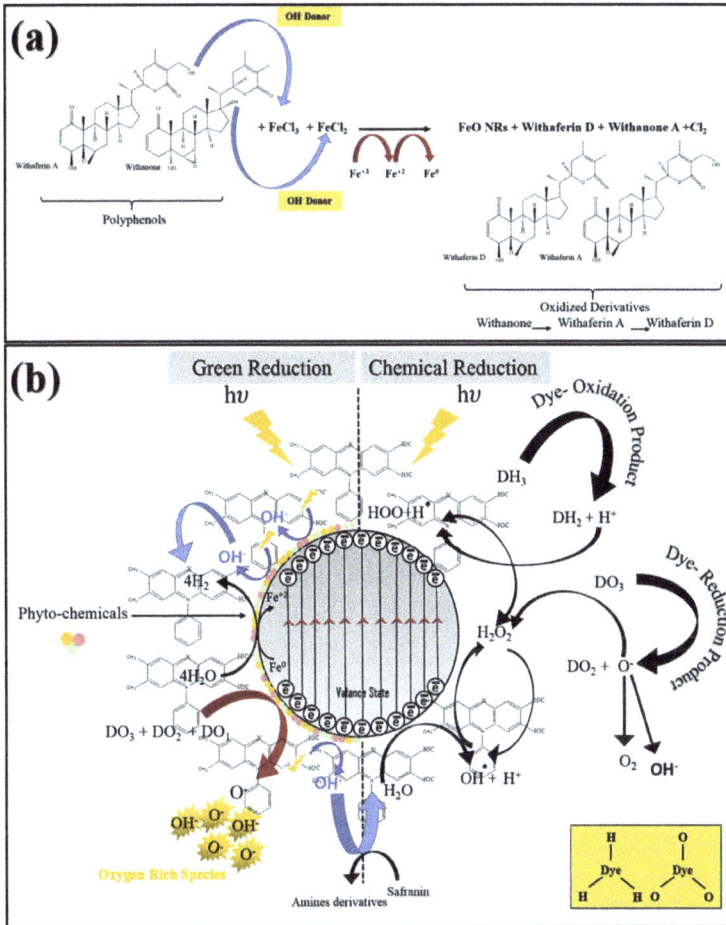

Fig. 7 *(a)*

Illustrates reduction and formation of iron oxides nanorods using Wathiana coagulans extract (b) Photocatalytic safranin dye degradation mechanism by iron oxides nanorods synthesis by both biological and chemical method. Reproduced with permission from Ref. [99].

In addition, pomegranate (*Punica granatum*) seeds extract-based Fe_2O_3 were fabricated through the green route, and the particle size was found to be in the range of 25-55 nm. The LCMS/MS was performed and identified the present biomolecule in the pomegranate seeds extract. The synthesized Fe_2O_3 was used for photocatalytic degradation of reactive blue (95.08%) [100]. Recently, biosynthesis of *Ruellia tuberosa* leaf extract-based iron oxide nanoparticles was reported by S. Vasantharaj et al. [101]. The 80% crystal violet dye degradation was achieved using the synthesized FeO NPs under solar irradiation.

Further, investigated the photocatalytic removal of cationic textile dye waste using *Kappaphycus alvarezii* plant extract mediated Fe_3O_4 nanoparticles [102]. Similarly, I. Fatimah et al. [103] have found bromophenol blue degradation was evaluated under UV and visible light using *Parkia speciosa Hassk* pod extract-based magnetic nanoparticles. However, their major disadvantage is low visible light utility and long-term stability. Synthesis of iron oxide nanoparticles utilizing the fruit extract of *Cynometra ramiflora* was reported. The nanoparticles' surface area ($107.97 \ m^2/g$) was determined by BET analysis of the MB dye degradation under sunlight [104]. Thus, further research is needed to effectively utilize these photocatalysts in the visible light region.

9.   Bio-mediated other metal oxide nanoparticles and its photocatalytic degradation towards toxic chemicals

Copper oxide nanoparticles are usually engaged for environmental treatment due to their more significant properties. For example, J. Singh et al. [105] reported the biogenic synthesis of copper oxide nanoparticles using *Psidium guajava* leaf extract. The formation of particles is spherical, with an average size of 2-6 nm and a surface area of $52.6 \ m^2/g$. The CuO NPs displayed an excellent degradation efficiency towards Nile blue (93% at 120 min) and reactive yellow 160 (81% at 120 min). The apparent rate constants of NB and RY160 were 0.023 and $0.014 \ min^{-1}$, respectively. Even after five consecutive cycles, photocatalytic dye degradation was found. The photocatalytic degradation of Coomassie brilliant blue R-250 using the green synthesis of copper oxide nanoparticles synthesized from *Carica papaya* leaves extract [106]. Recently, *Annona muricata* leaf extract-based low-cost CuO NPs via facile, non-toxic, and efficient route. The CuO NPs show the better photocatalytic ability for reactive red 120 (90% reaction time 60 min) and methyl orange (95% reaction time 60 min) degradation [107]. In addition, *Ananas comosus* leaf extract mediated nickel oxide nanoparticles were synthesized and characterized by various sophisticated analytical techniques. Finally, NiO NPs are utilized to degrade polyethylene film [108]. R. Li et al. [109] reported *Lippia citriodora* extract mediated synthesis Ag_2O NPs exhibits good photocatalytic degradation of acid orange 8 under UV light irradiation. In addition, Ag_2O with an excellent photocatalyst prepared using *Tridax* as a fuel reported by B.N. Rashmi et al. [110]. Recently, a green route for SnO_2 nanoparticles synthesis from *Vitex agnus-castus* fruit. They were utilized for rhodamine B dye mineralization under UV irradiation. The dye degradation efficiency was measured at 91.7% within 190 minutes [111]. Furthermore, R. Magudieshwaran et al. [112] reported *Jatropha curcas* plant extract mediated. The green route did CeO_2 nanoparticles and, the synthesized photocatalysis shows better acetaldehyde degradation performance. A. Angel Ezhilarasi et al. [113] have reported *Aegle marmelos* leaf extract via NiO nanoparticles used for the evaluating photodegradation of 4-chlorophenol. A *Brassica rapa* leaves extract-based functionalized cadmium tungstate (f-$CdWO_4$) nanoparticles were synthesized and shown in Fig. 8. The rod-shaped f-$CdWO_4$ nanoparticles and an average breadth of 27 nm and length 120 nm were found from TEM analysis. Finally, adsorption and photocatalytic degradation of toxic Bismarck brown R dye from aqueous solution using f-$CdWO_4$ under sunlight irradiation as shown in Fig. 8 [114].

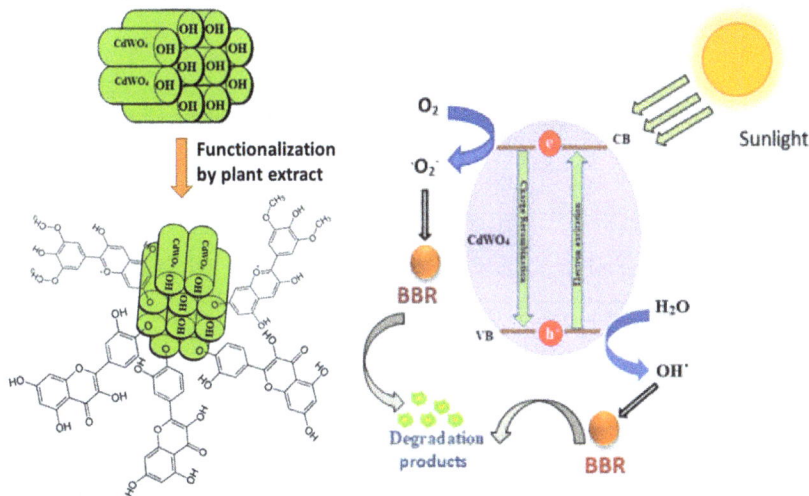

Fig. 8 Green functionalization of CdWO₄ nanoparticles and possible mechanism for degradation of toxic Bismarck brown R dye under sun light irradiation. Reproduced with permission from Ref. [114].

In addition, a green route synthesized SnO_2 nanoparticles using jujube fruit and investigated the photocatalytic activity for the degradation of eriochrome black-T (83%) and MB (90%) under direct sunlight. The photocatalyst was reused four times without any significant activity loss [115]. A.K.H. Bashir et al. [116] reported *Persea americana* seeds to extract NiO nanoparticles synthesized via an environmentally friendly green method. Used as photocatalysts in the free cyanide photodegradation. From cobalt nitrate hexahydrate, Cobalt-oxide nanoparticles were fabricated using *Punica granatum* peel extract and the size of 40-80 nm with a highly uniform shape. The 78.45% Brilliant Orange 3R dye degradation was achieved using synthesized NPs [117]. The detailed comparison is tabulated in Table 2.

Table 2. *Green synthesis of metal oxide nanoparticles and its photocatalytic degradation of toxic chemicals.*

Source	Metal oxide Nanoparticles	Size	Shape/morphology	Photocatalytic degradation	Ref.
Quince seed Mucilage	ZnO NPs	25 nm	-	methylene blue	80
Trianthema portulacastrum extract	ZnO NPs	25–90 nm	spherical	Synozol Navy Blue	81
Pithecellobium dulce	ZnO NPs	30 nm	spherical	Methylene blue	82
Calliandra haematocephala	ZnO NPs	-	flower-shape	methylene blue	85
Scutellaria baicalensis	ZnO NPs	50 nm	sphere-like structure	methylene blue	86
Withania coagulans	FeO NRs	16 ± 2 nm	spindle shape	safranin dye	99
Punica granatum seeds extract	Fe_2O_3 NPs	25–55 nm	spherical	reactive blue	100
Ruellia tuberosa leaf extract	FeO NPs	20–80 nm	rod	crystal violet	101
Kappaphycus alvarezii plant extract	Fe_3O_4 NPs	10 to 30 nm	-	hazardous textile waste	102
Psidium guajava leaf extract	CuO NPs	2–6 nm	spherical	Nile blue and reactive yellow 160	105
Carica papaya leaves extract	CuO NPs	140 nm	rod	brilliant blue R-250	106
Annona muricata leaf extract	CuO NPs	30–40 nm	spherical	reactive red 120 and methyl orange	107
Ananas comosus leaf extract	NiO NPs	1.42 ± 1.76 nm	-	polyethylene films	108
Lippia citriodora extract	Ag_2O NPs	20 nm	spherical	acid orange 8	109
Vitex agnus-castus fruit	SnO_2 NPs	4 to 13nm	spherical	rhodamine B	111
Jatropha curcas plant extract	CeO_2 NPs	3–5 nm	-	acetaldehyde	112
Aegle marmelos leaf extract	NiO NPs	8–10 nm	spherical	4-chlorophenol	113
Persea americana seeds	NiO NPs	11 nm	irregular shapes	free cyanide	116

Bioinspired Nanomaterials for Energy and Environmental Applications   Materials Research Forum LLC
Materials Research Foundations **121** (2022) 83-116   https://doi.org/10.21741/9781644901830-3

10. Bio-mediated metal oxide based nanocomposites for photocatalytic degradation of toxic chemicals

Increasing need for environmentally benign technologies, single and composite semiconductor metal oxides were synthesized. This greener approach has received growing attention because of its potential for excellent removal of toxic chemicals in wastewater. For instance, P. Panchal et al. [118] obtained a green synthesis of ZnO/MgO nanocomposites using *Ricinus Communis L*. The ZnO/MgO nanocomposites showed good photocatalytic activity towards MB. Compared to pure MgO and ZnO. RGO/NiO nanocomposites have been synthesized using *Psidium guajava L* (guava) leaf extract. The RGO/NiO nanocomposites (93%) show better degradation activity compared to NiO (65%) and RGO (negligible). The nanocomposites have three advantages (1) from NiO to RGO, a sufficient charge transfer takes place, (2) reduce electron-hole pair recombination, and (3) increase the dye degrade rate due to some greater extent [119]. Biosynthesized SnO_2 quantum dots (QDs) using Aparajitha (*Clitoria ternatea*) flower extract prepared. Used as effective photocatalysts for the rhodamine B degradation under UV light (Fig. 9) and optimized several parameters for the degradation of RhB. Likewise, a possible RhB photodegradation mechanism was investigated and proposed [120].

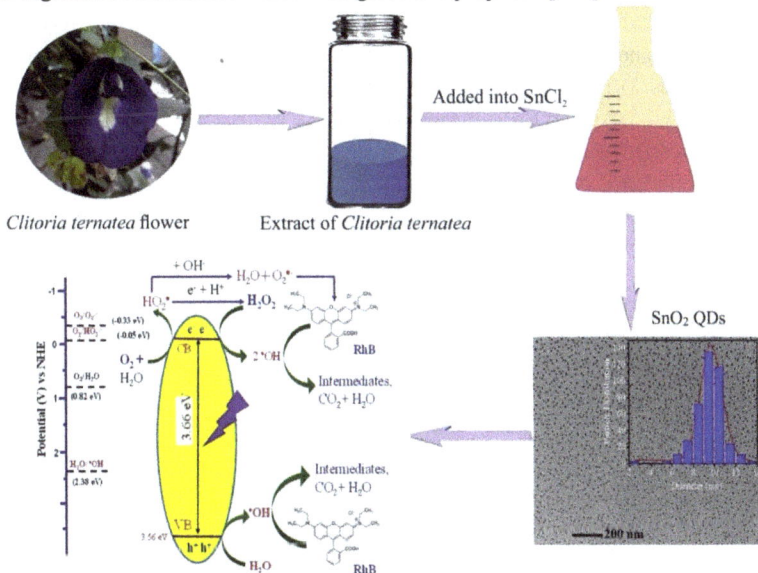

Biosynthesis of SnO₂ and its photocatalytic charge transfer mechanism towards RhB under UVB light irradiation. Reproduced with permission from Ref. [120].

Bioinspired Nanomaterials for Energy and Environmental Applications   Materials Research Forum LLC
Materials Research Foundations **121** (2022) 83-116   https://doi.org/10.21741/9781644901830-3

Furthermore, Azadirachta indica leaf extracts mediated Ag-Mo/CuO nanoparticles were prepared by K. Rajendran et al. [121] and investigated MB dye degradation. The Ag/ZnO-MMT nanomaterials were done using *urtica dioica* leaf extract, and it is utilized for the photo discoloration MB under visible light [122]. Mn/Mg-doped ZnO nanocomposite was synthesized by *Ocimum tenuiflorum* aqueous leaf extract and reported by R. Subbiah et al. [123]. Showes the highest photocatalytic activity towards MB. CuO and Ni@Fe$_3$O$_4$ nanoparticles were synthesized using *Euphorbia maculata* extract and used as photocatalysts for the degradation of methylene blue (MB), congo red (CR), and Rhodamine B (RhB). After four cycles, the photocatalyst efficiency did not observe any significant change [124].

11.   Microorganism mediated nanomaterials for photocatalytic degradation of toxic chemicals

In the past few decades, microorganism-based nanomaterials have gained attention due to their benefits over conventional chemical synthesis. Biosynthesis of titanium dioxide nanoparticles using *Bacillus amyloliquefaciens* culture prepared. Moreover, used for photocatalytic degradation of Reactive Red 31 dye under UV radiation [125]. R. M. Tripathi et al. [126] demonstrate a biosynthesis using *Bacillus licheniformis* MTCC 9555 mediated ZnO nanoflowers assessed for MB dye degradation under UV-light. Three repeated cycles of ZnO nanoflowers show good photostability. *Rhodotorula mucilaginosa* PA-1 mediated CdSe quantum dots were synthesized. They exhibited great photocatalytic activity towards malachite green (MG). They found the photocatalytic degradation efficiency of 86.5 % and 94.28 % and first-order rate constant (k) under ultraviolet and visible light found to be 0.0327 min−1 and 0.0397 min−1, respectively (Fig. 10 a-d). The GC-MS was used to analyze MG degradation intermediate products, and the result is shown in (Fig. 10e) [127]. Thus, further research is needed to utilize these photocatalysts effectively.

Bioinspired Nanomaterials for Energy and Environmental Applications   Materials Research Forum LLC
Materials Research Foundations **121** (2022) 83-116   https://doi.org/10.21741/9781644901830-3

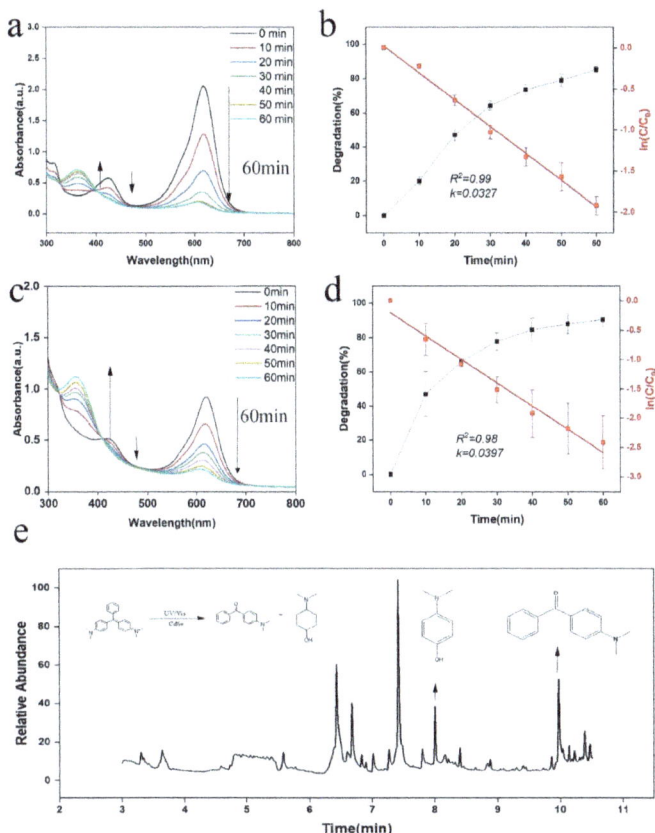

Fig. 10. The photocatalytic degradation of MG under UV-light (a) and Vis-light (c) using CdSe QDs. The photocatalyst degradation efficiency and first-order kinetics under the irradiation of UV-light (b) and Vis-light (d). The degradation products of MG were analyzed by GC-MS (e). Reproduced with permission from Ref. [127].

Conclusion and future outlook

This chapter briefly summarizes the recent bio-mediated nanomaterials preparation. The mechanism and photocatalytic degradation of toxic chemicals over the past few years. As discussed previously, various methodologies have been employed to synthesize nanomaterials and utilize multiple applications, especially in photocatalysis. Plant-mediated biosynthesis has some advantages: easy availability, cost-effective, environmentally friendly, and suitable for large-scale synthesis. The formation mechanism of nanoparticles, biological agents (stem, leaf, bark, and fruit), and metal ion precursors leading to the formation of nanoparticles with different morphology are not fully assumed clearly. Here we discussed up-to-date green-synthesized nanomaterials. The environmental remediation applications through photocatalysis. A more active biomolecule is needed for reducing and stabilizing synthesized nanomaterials. Also, nanoparticles are synthesized by utilizing bacteria, fungi, yeast. The utilization of photocatalytic application of microorganism-mediated synthesized nanomaterials is relatively low. Thus, varying microorganism and their substrate to photocatalytic degradation of toxic chemicals need further research in the future. We expected bio-mediated nanomaterials will make significant space and progress in these areas.

Acknowledgments

V.B. and K.S. would like to thank the Japan Society for the Promotion of Science (JSPS) for providing a postdoctoral fellowship for foreign researchers (P19393) and the research grant (JP19F19393).

References

[1] C. Santhosh, V. Velmurugan, G. Jacob, S.K. Jeong, A.N. Grace, A. Bhatnagar, Role of nanomaterials in water treatment applications: A review, Chem. Eng. J., 306 (2016) 1116-1137. https://doi.org/10.1016/j.cej.2016.08.053

[2] W. Zhang, M. Wang, W. Zhao, B. Wang, Magnetic composite photocatalyst $ZnFe_2O_4/BiVO_4$: synthesis, characterization, and visible-light photocatalytic activity, Dalton Trans., 42 (2013) 15464-15474. https://doi.org/10.1039/c3dt52068d

[3] S. Yi Lee, S.J. Park, TiO_2 photocatalyst for water treatment applications, J. Ind. Eng. Chem. 19 (2013) 1761-1769. https://doi.org/10.1016/j.jiec.2013.07.012

[4] A. Malathi, J. Madhavan, M. Ashokkumar, P. Arunachalam, A review on $BiVO_4$ photocatalyst: Activity enhancement methods for solar photocatalytic applications, Appl. Catal. A Gen., 555 (2018) 47-74. https://doi.org/10.1016/j.apcata.2018.02.010

[5] B. Vellaichamy, P. Periakaruppan, Ag nanoshell catalyzed dedying of industrial effluents, RSC Adv., 6 (2016) 31653-31660. https://doi.org/10.1039/C6RA02937J

[6] B. Vellaichamy, P. Periakaruppan, Synergistic combination of a novel metal-free mesoporous band-gap-modified carbon nitride grafted polyaniline nanocomposite for

decontamination of refractory pollutant, Ind. Eng. Chem. Res., 57 (2018) 6684-6695. https://doi.org/10.1021/acs.iecr.8b01098

[7]   B. Vellaichamy, P. Periakaruppan, R. Arumugam, K. Sellamuthu, B. Nagulan, A novel photocatalytically active mesoporous metal-free PPy grafted MWCNT nanocomposite, J. Colloid Interf. Sci., 514 (2018) 376-385. https://doi.org/10.1016/j.jcis.2017.12.046

[8]   S.K. Ponnaiah, P. Periakaruppan, B. Vellaichamy, B. Nagulan, Efficacious separation of electron-hole pairs in CeO_2-Al_2O_3 nanoparticles embedded GO heterojunction for robust visible-light driven dye degradation, J. Colloid Interf. Sci., 512 (2018) 219-230. https://doi.org/10.1016/j.jcis.2017.10.058

[9]   A.M. Awad, R. Jalab, A. Benamor, M.S. Nasser, M.M.B. Abbad, M.E. Naas, A.W. Mohammad, Adsorption of organic pollutants by nanomaterial-based adsorbents: An overview, J. Mol. Liq., 301 (2020) 112335. https://doi.org/10.1016/j.molliq.2019.112335

[10]   L. Zheng, Y. Jiao, H. Zhong, C. Zhang, J. Wang, Y. Wei, Insight into the magnetic lime coagulation-membrane distillation process for desulfurization wastewater treatment: From pollutant removal feature to membrane fouling, J. Hazard. Mater., 391 (2020) 122202. https://doi.org/10.1016/j.jhazmat.2020.122202

[11]   Q. Yang, Y. Ma, F. Chen, F. Yao, J. Sun, S. Wang, K. Yi, L. Hou, X. Li, D. Wang, Recent advances in photo-activated sulfate radical-advanced oxidation process (SR-AOP) for refractory organic pollutants removal in water, Chem. Eng. J., 378 (2019) 122149. https://doi.org/10.1016/j.cej.2019.122149

[12]   U. Baig, A. Matin, M.A. Gondal, S.M. Zubair, Facile fabrication of superhydrophobic, superoleophilic photocatalytic membrane for efficient oil-water separation and removal of hazardous organic pollutants, J. Clean. Prod., 208 (2019) 904-915. https://doi.org/10.1016/j.jclepro.2018.10.079

[13]   B.G. Ryu, J. Kim, J.I. Han, K. Kim, D. Kim, B.K. Seo, C.M. Kang, J.W. yang, Evaluation of an electro-flotation-oxidation process for harvesting bio-flocculated algal biomass and simultaneous treatment of residual pollutants in coke wastewater following an algal-bacterial process, Algal Res., 31 (2018) 497-505. https://doi.org/10.1016/j.algal.2017.06.012

[14]   E. Vaudevire, F. Radmanesh, A. Kolkman, D. Vughs, E. Cornelissen, J. Post, W.V.D. Meer, Fate and removal of trace pollutants from an anion exchange spent brine during the recovery process of natural organic matter and salts, Water Res., 154 (2019) 34-44. https://doi.org/10.1016/j.watres.2019.01.042

[15] X.H. Lin, S.F.Y. Li, Impact of the spatial distribution of sulfate species on the activities of SO_4^{2-}/TiO_2 photocatalysts for the degradation of organic pollutants in reverse osmosis concentrate, Appl. Catal. B Environ., 170-171 (2015) 263-272. https://doi.org/10.1016/j.apcatb.2015.02.001

[16] D. Yu, J. Cui, X. Li, H. Zhang, Y. Pei, Electrochemical treatment of organic pollutants in landfill leachate using a three-dimensional electrode system, Chemosphere, 243 (2020) 125438. https://doi.org/10.1016/j.chemosphere.2019.125438

[17] B. Vellaichamy, P. Periakaruppan, B. Nagulan, Reduction of Cr^{6+} from wastewater using a novel in situ-synthesized $PANI/MnO_2/TiO_2$ nanocomposite: renewable, selective, stable, and synergistic catalysis, ACS Sustain. Chem. Eng., 5 (2017) 9313-9324. https://doi.org/10.1021/acssuschemeng.7b02324

[18] B. Vellaichamy, P. Periakaruppan, A facile in situ synthesis of highly active and reusable ternary Ag-PPy-GO nanocomposite for catalytic oxidation of hydroquinone in aqueous solution, J. Catal., 344 (2016) 795-805. https://doi.org/10.1016/j.jcat.2016.08.010

[19] B. Vellaichamy, P. Periakaruppan, Catalytic hydrogenation performance of an in situ assembled $Au@g\text{-}C_3N_4\text{-}PANI$ nanoblend: synergistic inter-constituent interactions boost the catalysis, New J. Chem., 41 (2017) 7123-7132. https://doi.org/10.1039/C7NJ01085K

[20] B. Vellaichamy, P. Periakaruppan, Silver nanoparticle-embedded RGO-nanosponge for superior catalytic activity towards 4-nitrophenol reduction, RSC Adv., 6 (2016) 88837-88845. https://doi.org/10.1039/C6RA19834A

[21] V. Balakumar, H. Kim, J.W. Ryu, R. Manivannan, Y.A. Son, Uniform assembly of gold nanoparticles on S-doped $g\text{-}C_3N_4$ nanocomposite for effective conversion of 4-nitrophenol by catalytic reduction, J. Mater. Sci. Technol., 40 (2020) 176-184. https://doi.org/10.1016/j.jmst.2019.08.031

[22] V. Balakumar, H. Kim, R. Manivannan, H. Kim, J.W. Ryu, G. Heo, Y.A. Son, Ultrasound-assisted method to improve the structure of $CeO_2@polypyrrole$ core-shell nanosphere and its photocatalytic reduction of hazardous Cr^{6+}, Ultrason. Sonochem., 59 (2019) 104738. https://doi.org/10.1016/j.ultsonch.2019.104738

[23] K.S. Kumar, B. Vellaichamy, T. Paulmony, Visible Light Active Metal-Free Photocatalysis: N-Doped Graphene Covalently Grafted with $g\text{-}C_3N_4$ for Highly Robust Degradation of Methyl Orange, Solid State Sci., 94 (2019) 99-105. https://doi.org/10.1016/j.solidstatesciences.2019.06.003

[24] Z. Xiong, Z. Xu, Y. Li, L. Dong, J. Wang, J. Zhao, X. Chen, Y. Zhao, H. Zhao, J. Zhang, Incorporating highly dispersed and stable Cu^+ into TiO_2 lattice for enhanced photocatalytic CO_2 reduction with water, Appl. Surf. Sci., 507 (2020) 145095. https://doi.org/10.1016/j.apsusc.2019.145095

[25] D. Zhong, W. Liu, P. Tan, A. Zhu, Y. Liu, X. Xiong, J. Pan, Insights into the synergy effect of anisotropic {001} and {230}facets of $BaTiO_3$ nanocubes sensitized with CdSe quantum dots for photocatalytic water reduction, Appl. Catal. B Environ., 227 (2018) 1-12. https://doi.org/10.1016/j.apcatb.2018.01.009

[26] T. Billo, I. Shown, A.K. Anbalagan, T.A. Effendi, A. Sabbah, F.Y. Fu, C.M. Chu, W.Y. Woon, R.S. Chen, C.H. Lee, K.H. Chen, L.C. Chen, A mechanistic study of molecular CO_2 interaction and adsorption on carbon implanted SnS_2 thin film for photocatalytic CO_2 reduction activity, Nano Energy, 72 (2020) 104717. https://doi.org/10.1016/j.nanoen.2020.104717

[27] Q. Guo, L. Fu, T. Yan, W. Tian, D. Ma, J. Li, Y. Jiang, X. Wang, Improved photocatalytic activity of porous ZnO nanosheets by thermal deposition graphene-like g-C_3N_4 for CO_2 reduction with H_2O vapor, Appl. Surf. Sci., 509 (2020) 144773. https://doi.org/10.1016/j.apsusc.2019.144773

[28] X.F. Chen, S.C. Kou, C.S. Poon, Rheological behaviour, mechanical performance, and NOx removal of photocatalytic mortar with combined clay brick sands-based and recycled glass-based nano-TiO_2 composite photocatalysts, Constr. Build. Mater., 240 (2020) 117698. https://doi.org/10.1016/j.conbuildmat.2019.117698

[29] J.M. Cordero, R. Hingorani, J. Relinque, M. Grande, R. Borge, A. Narros, M. Castellote, NOx removal efficiency of urban photocatalytic pavements at pilot scale, Sci. Total Environ., 719 (2020) 137459. https://doi.org/10.1016/j.scitotenv.2020.137459

[30] C. Li, Y. Xu, W. Tu, G. Chen, R. Xu, Metal-free photocatalysts for various applications in energy conversion and environmental purification, Green Chem., 19 (2017) 882-899. https://doi.org/10.1039/C6GC02856J

[31] W. Xue, D. Huang, X. Wen, S. Chen, M. Cheng, R. Deng, B. Li, Y. Yang, X. Liu, Silver-based semiconductor Z-scheme photocatalytic systems for environmental purification, J. Hazard. Mater., 390 (2020) 122128. https://doi.org/10.1016/j.jhazmat.2020.122128

[32] P. Singh, P. Shandilya, P. Raizada, A. Sudhaik, A.R. Sani, A.H. Bandegharaei, Review on various strategies for enhancing photocatalytic activity of graphene based nanocomposites for water purification, Arab. J. Chem., 13 (2020) 3498-3520. https://doi.org/10.1016/j.arabjc.2018.12.001

[33] X. Zhang, J. Wang, X.X. Dong, Y.K. Lv, Functionalized metal-organic frameworks for photocatalytic degradation of organic pollutants in environment, Chemosphere, 242 (2020) 125144. https://doi.org/10.1016/j.chemosphere.2019.125144

[34] R. Gusain, K. Gupta, P. Joshi, O.P. Khatri, Adsorptive removal and photocatalytic degradation of organic pollutants using metal oxides and their composites: A comprehensive review, Adv. Colloid Interfac., 272 (2019) 102009. https://doi.org/10.1016/j.cis.2019.102009

[35] X. Sun, Z. Liu, H. Yu, Z. Zheng, D. Zeng, Facile synthesis of $BiFeO_3$ nanoparticles by modified microwave-assisted hydrothermal method as visible light driven photocatalysts, Mater. Lett., 219 (2018) 225-228. https://doi.org/10.1016/j.matlet.2018.02.052

[36] E. Keles, M. Yildirim, T. Ozturk, O.A. Yildirim, Hydrothermally synthesized UV light active zinc stannate:tin oxide ($ZTO:SnO_2$) nanocomposite photocatalysts for photocatalytic applications, Mat. Sci. Semicon. Proc., 110 (2020) 104959. https://doi.org/10.1016/j.mssp.2020.104959

[37] Z. Xing, Y. Chen, C. Liu, J. Yang, J. Xu, Y. Situ, H. Huang, Synthesis of core-shell ZnO/oxygen doped g-C_3N_4 visible light driven photocatalyst via hydrothermal method, J. Alloy. Compd., 708 (2017) 853-861. https://doi.org/10.1016/j.jallcom.2016.11.295

[38] P. Zhang, S. Yin, V. Petrykin, M. Kakihana, T. Sato, Preparation of high performance fibrous titania photocatalysts by the solvothermal reaction of protonated form of tetratitanate, J. Mol. Catal. A Chem., 309 (2009) 50-56. https://doi.org/10.1016/j.molcata.2009.04.014

[39] T. Chankhanittha, V. Somaudon, J. Watcharakitti, V. Piyavarakorn, S, Nanan Performance of solvothermally grown Bi_2MoO_6 photocatalyst toward degradation of organic azo dyes and fluoroquinolone antibiotics, Mater. Lett., 258 (2020) 126764. https://doi.org/10.1016/j.matlet.2019.126764

[40] Y. Li, Y. Song, Z. Ma, S. Niu, J. Li, N. Li, Synthesis of phosphonic acid silver-graphene oxide nanomaterials with photocatalytic activity through ultrasonic-assisted method, Ultrason. Sonochem., 44 (2018) 106-114. https://doi.org/10.1016/j.ultsonch.2018.02.026

[41] B. Liang, C. Sun, D. Han, W. Zhang, R. Zhang, Y. Zhang, Y. Liu, Effect of annealing temperature on structure and photocatalytic efficiency of SnO microspheres synthesized by ultrasonic reaction method, Ceram. Int., 44 (2018) 23334-23338. https://doi.org/10.1016/j.ceramint.2018.09.112

[42] A. Manohar, C. Krishnamoorthi, Magnetic and photocatalytic studies on $Zn_{1-x}Mg_xFe_2O_4$ nanocolloids synthesized by solvothermal reflux method, J. Photoch. Photobio. B, 177 (2017) 95-104. https://doi.org/10.1016/j.jphotobiol.2017.10.009

[43] J. Liang, J. Wang, K. Song, X. Wang, K. Yu, C. Liang, Enhanced photocatalytic activities of Nd-doped TiO_2 under visible light using a facile sol-gel method, J. Rare Earth., 38 (2020) 148-156. https://doi.org/10.1016/j.jre.2019.07.008

[44] Y. Zhai, Y. Yin, W. Zhang, Y. Han, X. Zhao, X. Liu, One-step synthesis of butterfly-like SnC_2O_4 by simple precipitation method and photocatalytic performance, Mat. Sci. Semicon. Proc., 56 (2016) 83-89. https://doi.org/10.1016/j.mssp.2016.07.010

[45] M. Zhang, Y. Li, Q. Wang, R. Jin, H. Xu, S. Gao, Effect of different reductants on the composition and photocatalytic performances of $Ag/AgIO_3$ hybrids prepared by in-situ reduction method, Inorg. Chem. Commun., 115 (2020) 107876. https://doi.org/10.1016/j.inoche.2020.107876

[46] M. Bandeira, M. Giovanela, M.R. Ely, D.M. Devine, J.D.S. Crespo, Green synthesis of zinc oxide nanoparticles: A review of the synthesis methodology and mechanism of formation, Sustain. Chem. Pharm., 15 (2020) 100223. https://doi.org/10.1016/j.scp.2020.100223

[47] V. Hoseinpour, N. Ghaemi, Green synthesis of manganese nanoparticles: Applications and future perspective-A review, J. Photoch. Photobio. B, 189 (2018) 234-243. https://doi.org/10.1016/j.jphotobiol.2018.10.022

[48] A. Singh, P.K. Gautam, A. Verma, V. Singh, P.M. Shivapriya, S. Shivalkar, A.K. Sahoo, S.K. Samanta, Green synthesis of metallic nanoparticles as effective alternatives to treat antibiotics resistant bacterial infections: A review, Biotechnol. Rep., 25 (2020) e00427. https://doi.org/10.1016/j.btre.2020.e00427

[49] M.N. Owaid, Green synthesis of silver nanoparticles by *Pleurotus* (oyster mushroom) and their bioactivity: Review, Environ. Nanotechnol. Monit. Manag., 12 (2019) 100256. https://doi.org/10.1016/j.enmm.2019.100256

[50] S. Ahmed, Annu, S.A. Chaudhry, S. Ikram, A review on biogenic synthesis of ZnO nanoparticles using plant extracts and microbes: A prospect towards green chemistry, J. Photoch. Photobio. B, 166 (2017) 272-284. https://doi.org/10.1016/j.jphotobiol.2016.12.011

[51] V.K.M. Katta, R.S. Dubey, Green synthesis of silver nanoparticles using *Tagetes erecta* plant and investigation of their structural, optical, chemical and morphological properties, Mater. Today Proc., (2020). https://doi.org/10.1016/j.matpr.2020.02.809

[52] U.B. Manik, A. Nande, S. Raut, S.J. Dhoble, Green synthesis of silver nanoparticles using plant leaf extraction of *Artocarpus heterophylus* and *Azadirachta indica*, Results Mater., 6 (2020) 100086. https://doi.org/10.1016/j.rinma.2020.100086

[53] W. Ahmad, D. Kalra, Green synthesis, characterization and anti-microbial activities of ZnO nanoparticles using *Euphorbia hirta* leaf extract, J. King Saud Univ. Sci., (2020). https://doi.org/10.1016/j.jksus.2020.03.014

[54] P. Narasaiah, B.K. Mandal, N.C. Sarada, Green synthesis of Pd NPs from *Pimpinella tirupatiensis* plant extract and their application in photocatalytic activity dye degradation, IOP Conf. Ser. Mat. Sci., 263 (2017) 022013. https://doi.org/10.1088/1757-899X/263/2/022013

[55] E. Haritha, S.M. Roopan, G. Madhavi, G. Elango, N.A.A. Dhabi, M.V. Arasu, Environmental friendly synthesis of palladium nanoparticles and its photocatalytic activity against diazo dye to sustain the natural source, J. Clust. Sci., 28 (2017) 1225-1236. https://doi.org/10.1007/s10876-016-1136-2

[56] A.A. Olajire, A.A. Mohammed, Green synthesis of palladium nanoparticles using *Ananas comosus* leaf extract for solid-phase photocatalytic degradation of low density polyethylene film, J. Environ. Chem. Eng., 7 (2019) 103270. https://doi.org/10.1016/j.jece.2019.103270

[57] P. Kumar, M. Govindaraju, S. Senthamilselvi, K. Premkumar, Photocatalytic degradation of methyl orange dye using silver (Ag) nanoparticles synthesized from *Ulva lactuca*, Colloid. Surface. B, 103 (2013) 658-661. https://doi.org/10.1016/j.colsurfb.2012.11.022

[58] B. Kumar, K. Smita, L. Cumbal, A. Debut, Sacha inchi (*Plukenetia volubilis* L.) oil for one pot synthesis of silver nanocatalyst: An ecofriendly approach, Ind. Crop. Prod., 58 (2014) 238-243. https://doi.org/10.1016/j.indcrop.2014.04.021

[59] D. Baruah, R.N.S. Yadav, A. Yadav, A.M. Das, *Alpinia nigra* fruits mediated synthesis of silver nanoparticles and their antimicrobial and photocatalytic activities, J. Photoch. Photobio. B, 201 (2019) 111649. https://doi.org/10.1016/j.jphotobiol.2019.111649

[60] D. Baruah, M. Goswami, R.N.S. Yadav, A. Yadav, A.M. Das, Biogenic synthesis of gold nanoparticles and their application in photocatalytic degradation of toxic dyes, J. Photoch. Photobio. B, 186 (2018) 51-58. https://doi.org/10.1016/j.jphotobiol.2018.07.002

[61] S.P. Vinay, N. Chandrasekhar, Facile green chemistry synthesis of Ag nanoparticles using *Areca catechu* extracts for the antimicrobial activity and

photocatalytic degradation of methylene blue dye, Mater. Today Proc., 9 (2019) 499-505. https://doi.org/10.1016/j.matpr.2018.10.368

[62] J.F. Li, Y.C. Liu, M. Chokkalingam, E.J. Rupa, R. Mathiyalagan, J. Hurh, J.C. Ahn, J.K. Park, J.Y. Pu, D.C. Yang, Phytosynthesis of silver nanoparticles using rhizome extract of *Alpinia officinarum* and their photocatalytic removal of dye under UV and visible light irradiation, Optik, 208 (2020) 164521. https://doi.org/10.1016/j.ijleo.2020.164521

[63] B.C. Choudhary, D. Paul, T. Gupta, S.R. Tetgure, V.J. Garole, A.U. Borse, D.J. Garole, Photocatalytic reduction of organic pollutant under visible light by green route synthesized gold nanoparticles, J. Environ. Sci., 55 (2017) 236-246. https://doi.org/10.1016/j.jes.2016.05.044

[64] B. Paul, B. Bhuyan, D.D. Purkayastha, M. Dey, S.S. Dhar, Green synthesis of gold nanoparticles using *Pogestemon benghalensis (B) O. Ktz.* leaf extract and studies of their photocatalytic activity in degradation of methylene blue, Mater. Lett., 148 (2015) 37-40. https://doi.org/10.1016/j.matlet.2015.02.054

[65] G.A. Molina, R. R. Esparza, J.L.L. Miranda, A.R.H. Martinez, B.L.E. Sanchez, E.A.E. Pena, M. Estevez, Green synthesis of Ag nanoflowers using *Kalanchoe daigremontiana* extract for enhanced photocatalytic and antibacterial activities, Colloid. Surface. B, 180 (2019) 141-149. https://doi.org/10.1016/j.colsurfb.2019.04.044

[66] A. Ahmed, M. Usman, B. Yu, X. Ding, Q. Peng, Y. Shen, H. Cong, Efficient photocatalytic degradation of toxic Alizarin yellow R dye from industrial wastewater using biosynthesized Fe nanoparticle and study of factors affecting the degradation rate, J. Photoch. Photobio. B, 202 (2020) 111682. https://doi.org/10.1016/j.jphotobiol.2019.111682

[67] I.A. Radini, N. Hasan, M.A. Malik, Z. Khan, Biosynthesis of iron nanoparticles using *Trigonella foenum-graecum* seed extract for photocatalytic methyl orange dye degradation and antibacterial applications, J. Photoch. Photobio. B, 183 (2018) 154-163. https://doi.org/10.1016/j.jphotobiol.2018.04.014

[68] U. Kamran, H.N. Bhatti, M. Iqbal, S. Jamil, M. Zahid, Biogenic synthesis, characterization and investigation of photocatalytic and antimicrobial activity of manganese nanoparticles synthesized from *Cinnamomum verum* bark extract, J. Mol. Struct., 1179 (2019) 532-539. https://doi.org/10.1016/j.molstruc.2018.11.006

[69] K. Roy, C.K. Ghosh, C.K. Sarkar, Degradation of toxic textile dyes and detection of hazardous Hg^{2+} by low-cost bioengineered copper nanoparticles synthesized using

Impatiens balsamina leaf extract, Mater. Res. Bull., 94 (2017) 257-262.
https://doi.org/10.1016/j.materresbull.2017.06.016

[70] I. Bibi, S. Kamal, A. Ahmed, M. Iqbal, S. Nouren, K. Jilani, N. Nazar, M. Amir, A. Abbas, S. Ata, F. Majid, Nickel nanoparticle synthesis using *Camellia Sinensis* as reducing and capping agent: Growth mechanism and photo-catalytic activity evaluation, Int. J. Biol. Macromol., 103 (2017) 783-790.
https://doi.org/10.1016/j.ijbiomac.2017.05.023

[71] S. Shams, A.U. Khan, Q. Yuan, W. Ahmad, Y. Wei, Z.U.H. Khan, S. Shams, A. Ahmad, A.U. Rahman, S. Ullah, Facile and eco-benign synthesis of Au@Fe_2O_3 nanocomposite: Efficient photocatalytic, antibacterial and antioxidant agent, J. Photoch. Photobio. B, 199 (2019) 111632. https://doi.org/10.1016/j.jphotobiol.2019.111632

[72] C. Karthik, N. Swathi, S.P. Prabha, D.G. Caroline, Green synthesized rGO-AgNP hybrid nanocomposite-An effective antibacterial adsorbent for photocatalytic removal of DB-14 dye from aqueous solution, J. Environ. Chem. Eng., 8 (2020) 103577.
https://doi.org/10.1016/j.jece.2019.103577

[73] P. Panchal, D.R. Paul, A. Sharma, P. Choudhary, P. Meena, S.P. Nehra, Biogenic mediated Ag/ZnO nanocomposites for photocatalytic and antibacterial activities towards disinfection of water, J. Colloid Interf. Sci., 563 (2020) 370-380.
https://doi.org/10.1016/j.jcis.2019.12.079

[74] K.S. Vizuete, B. Kumar, A.V. Vaca, A. Debut, L. Cumbal, Mortiño (*Vaccinium floribundum* Kunth) berry assisted green synthesis and photocatalytic performance of silver-graphene nanocomposite, J. Photoch. Photobio. A, 329 (2016) 273-279.
https://doi.org/10.1016/j.jphotochem.2016.06.030

[75] D. Ayodhya, G. Veerabhadram, Green synthesis of garlic extract stabilized Ag@CeO_2 composites for photocatalytic and sonocatalytic degradation of mixed dyes and antimicrobial studies, J. Mol. Struct., 1205 (2020) 127611.
https://doi.org/10.1016/j.molstruc.2019.127611

[76] N.L. Gavade, A.N. Kadam, S.B. Babar, A.D. Gophane, K.M. Garadkar, S.W. Lee, Biogenic synthesis of gold-anchored ZnO nanorods as photocatalyst for sunlight-induced degradation of dye effluent and its toxicity assessment, Ceram. Int., 46 (2020) 11317-11327. https://doi.org/10.1016/j.ceramint.2020.01.161

[77] S.H. Khan, B. Pathak, ZnO based photocatalytic degradation of persistent pesticides: A comprehensive review, Environ. Nanotechnol. Monit. Manag., DOI:doi.org/10.1016/ j.enmm.2020.100290 (2020).
https://doi.org/10.1016/j.enmm.2020.100290

[78] M Golmohammadi, M. Honarmand, S. Ghanbari, A green approach to synthesis of ZnO nanoparticles using jujube fruit extract and their application in photocatalytic degradation of organic dyes, Spectrochim. Acta A, 229 (2020) 117961. https://doi.org/10.1016/j.saa.2019.117961

[79] U. Singhal, R. Pendurthi, M. Khanuja, Prunus: A natural source for synthesis of zinc oxide nanoparticles towards photocatalytic and antibacterial applications, Mater. Today Proc., doi.org/10.1016/j.matpr.2020.01.606 (2020). https://doi.org/10.1016/j.matpr.2020.01.606

[80] S.M.T.H. Moghaddas, B. Elahi, V. Javanbakht, Biosynthesis of pure zinc oxide nanoparticles using Quince seed mucilage for photocatalytic dye degradation, J. Alloy. Compd., 821 (2020) 153519. https://doi.org/10.1016/j.jallcom.2019.153519

[81] Z.U.H. Khan, H.M. Sadiq, N.S. Shah, A.U. Khan, N. Muhammad, S.U. Hasan, K. Tahir, S.Z. Safi, F.U. Khan, M. Imran, N. Ahmad, F. Ullah, A. Ahmad, M. Sayed, M.S. Khalid, S.A. Qaisrani, M. Ali, A. Zakir, Greener synthesis of zinc oxide nanoparticles using *Trianthema portulacastrum* extract and evaluation of its photocatalytic and biological applications, J. Photoch. Photobio. B, 192 (2019) 147-157. https://doi.org/10.1016/j.jphotobiol.2019.01.013

[82] G. Madhumitha, J. Fowsiya, N. Gupta, A. Kumar, M. Singh, Green synthesis, characterization and antifungal and photocatalytic activity of *Pithecellobium dulce* peel-mediated ZnO nanoparticles, J. Phys. Chem. Solids, 127 (2019) 43-51. https://doi.org/10.1016/j.jpcs.2018.12.005

[83] Y. Zheng, L. Fu, F. Han, A. Wang, W. Cai, J. Yu, J. Yang, F. Peng, Green biosynthesis and characterization of zinc oxide nanoparticles using *Corymbia citriodora* leaf extract and their photocatalytic activity, Green Chem. Lett. Rev., 8 (2015) 59-63. https://doi.org/10.1080/17518253.2015.1075069

[84] S.S.M. Hassan, W.I.M.E. Azab, H.R. Ali, M.S.M. Mansour, Green synthesis and characterization of ZnO nanoparticles for photocatalytic degradation of anthracene, Adv. Nat. Sci. Nanosci., 6 (2015) 045012. https://doi.org/10.1088/2043-6262/6/4/045012

[85] R. Vinayagam, R. Selvaraj, P. Arivalagan, T. varadavenkatesan, Synthesis, characterization and photocatalytic dye degradation capability of *Calliandra haematocephala*-mediated zinc oxide nanoflowers, J. Photoch. Photobio. B, 203 (2020) 111760. https://doi.org/10.1016/j.jphotobiol.2019.111760

[86] L. Chen, I. Batjikh, J. Hurh, Y. Han, Y. Huo, H. Ali, J.F. Li, E.J. Rupa, J.C. Ahn, R. Mathiyalagan, D.C. Yang, Green synthesis of zinc oxide nanoparticles from root

extract of *Scutellaria baicalensis* and its photocatalytic degradation activity using methylene blue, Optik, 184 (2019) 324-329. https://doi.org/10.1016/j.ijleo.2019.03.051

[87] D. Suresh, P.C. Nethravathi, Udayabhanu, H. Rajanaika, H. Nagabhushana, S.C. Sharma, Green synthesis of multifunctional zinc oxide (ZnO) nanoparticles using *Cassia fistula* plant extract and their photodegradative, antioxidant and antibacterial activities, Mat. Sci. Semicon. Proc., 31 (2015) 446-454. https://doi.org/10.1016/j.mssp.2014.12.023

[88] J. Lu, H. Ali, J. Hurh, Y. Han, I. Batjikh, E.J. Rupa, G. Anandapadmanaban, J.K. Park, D.C. Yang, The assessment of photocatalytic activity of zinc oxide nanoparticles from the roots of *Codonopsis lanceolata* synthesized by one-pot green synthesis method, Optik, 184 (2019) 82-89. https://doi.org/10.1016/j.ijleo.2019.03.050

[89] B. Siripireddy, B.K. Mandal, Facile green synthesis of zinc oxide nanoparticles by *Eucalyptus globulus* and their photocatalytic and antioxidant activity, Adv. Powder Technol., 28 (2017) 785-797. https://doi.org/10.1016/j.apt.2016.11.026

[90] B.P. Kumar, M. Arthanareeswari, S. Devikala, M. Sridharan, J.A. Selvi, T.P. Malini, Green synthesis of zinc oxide nanoparticles using typha *latifolia*. L leaf extract for photocatalytic applications, Mater. Today Proc., 14 (2019) 332-337. https://doi.org/10.1016/j.matpr.2019.04.155

[91] M. Stan, A. Popa, D. Toloman, A. Dehelean, I. Lung, G. Katona, Enhanced photocatalytic degradation properties of zinc oxide nanoparticles synthesized by using plant extracts, Mat. Sci. Semicon. Proc., 39 (2015) 23-29. https://doi.org/10.1016/j.mssp.2015.04.038

[92] J. Lu, I. Batjikh, J. Hurh, Y. Han, H. Ali, R. Mathiyalagan, C. Ling, J.C. Ahn, D.C. Yang, Photocatalytic degradation of methylene blue using biosynthesized zinc oxide nanoparticles from bark extract of *Kalopanax septemlobus*, Optik, 182 (2019) 980-985. https://doi.org/10.1016/j.ijleo.2018.12.016

[93] Y.J. Shim, V. Soshnikova, G. Anandapadmanaban, R. Mathiyalagan, Z.E.J. Perez, J. Markus, Y.J. Kim, V.C. Aceituno, D.C. Yang, Zinc oxide nanoparticles synthesized by *Suaeda japonica* Makino and their photocatalytic degradation of methylene blue, Optik, 182 (2019) 1015-1020. https://doi.org/10.1016/j.ijleo.2018.11.144

[94] T. Bhuyan, K. Mishra, M. Khanuja, R. Prasad, A. Varma, Biosynthesis of zinc oxide nanoparticles from *Azadirachta indica* for antibacterial and photocatalytic applications, Mat. Sci. Semicon. Proc., 32 (2015) 55-61. https://doi.org/10.1016/j.mssp.2014.12.053

[95] S. Kanagasubbulakshmi, K. Kadirvelu, Green synthesis of iron oxide nanoparticles using *Lagenaria siceraria* and evaluation of its antimicrobial activity, Defence Life Sci. J., 2 (2017) 422-427. https://doi.org/10.14429/dlsj.2.12277

[96] K. Gottimukkala, R. Harika, D. Zamare, Green synthesis of iron nanoparticles using green tea leaves extract, J. Nanomed. Biotherapeut. Discov., 7 (2017) 1000151.

[97] S. Amutha, S. Sridhar, Green synthesis of magnetic iron oxide nanoparticle using leaves of *Glycosmis mauritiana* and their antibacterial activity against human pathogens, J. Innovat. Pharm. Biol. Sci., 5 (2018) 22-26.

[98] G. Jagathesan, P. Rajiv, Biosynthesis and characterization of iron oxide nanoparticles using *Eichhornia crassipes* leaf extract and assessing their antibacterial activity, Biocat. Agric. Biotechnol., 13 (2018) 90-94. https://doi.org/10.1016/j.bcab.2017.11.014

[99] S. Qasim, A. Zafer, M.S. Saif, Z. Ali, M. Nazar, M. Waqas, A.U. Haq, T. Tariq, S.G. Hassan, F. Iqbal, X.G. Shu, M. Hasan, Green synthesis of iron oxide nanorods using *Withania coagulans* extract improved photocatalytic degradation and antimicrobial activity, J. Photoch. Photobio. B, 204 (2020) 111784. https://doi.org/10.1016/j.jphotobiol.2020.111784

[100] I. Bibi, N. Nazar, S. Ata, M. Sultan, A. Ali, A. Abbas, K. Jilani, S. Kamal, F.M. Sarim, M.I. Khan, F. Jalal, M. Iqbal, Green synthesis of iron oxide nanoparticles using pomegranate seeds extract and photocatalytic activity evaluation for the degradation of textile dye, J. Mater. Res. Technol., 8 (2019) 6115-6124. https://doi.org/10.1016/j.jmrt.2019.10.006

[101] S. Vasantharaj, S. sathiyavimal, P. Senthilkumar, F.L. Oscar, A. Pugazhendhi, Biosynthesis of iron oxide nanoparticles using leaf extract of *Ruellia tuberosa*: Antimicrobial properties and their applications in photocatalytic degradation, J. Photoch. Photobio. B, 192 (2019) 74-82. https://doi.org/10.1016/j.jphotobiol.2018.12.025

[102] M.V. Arularasu, J. Devakumar, T.V. Rajendran, An innovative approach for green synthesis of iron oxide nanoparticles: Characterization and its photocatalytic activity, Polyhedron, 156 (2018) 279-290. https://doi.org/10.1016/j.poly.2018.09.036

[103] I. Fatimah, E.Z. Pratiwi, W.P. Wicaksono, Synthesis of magnetic nanoparticles using *Parkia speciosa Hassk* pod extract and photocatalytic activity for Bromophenol blue degradation, Egypt. J. Aquat. Res., 46 (2020) 35-40. https://doi.org/10.1016/j.ejar.2020.01.001

[104] S. Bishnoi, A. Kumar, R. Selvaraj, Facile synthesis of magnetic iron oxide nanoparticles using inedible *Cynometra ramiflora* fruit extract waste and their

photocatalytic degradation of methylene blue dye, Mater. Res. Bull., 97 (2018) 121-127. https://doi.org/10.1016/j.materresbull.2017.08.040

[105] J. Singh, V. Kumar, K.H. Kim, M. Rawat, Biogenic synthesis of copper oxide nanoparticles using plant extract and its prodigious potential for photocatalytic degradation of dyes, Environ. Res., 177 (2019) 108569. https://doi.org/10.1016/j.envres.2019.108569

[106] R. Sankar, P. Manikandan, V. Malarvizhi, T. Fathima, K.S. Shivashangari, V. Ravikumar, Green synthesis of colloidal copper oxide nanoparticles using *Carica papaya* and its application in photocatalytic dye degradation, Spectrochim. Acta A, 121 (2014) 746-750. https://doi.org/10.1016/j.saa.2013.12.020

[107] K. Sukumar, S. Arumugam, S. Thangaswamy, S. Balakrishnan, S. Chinnappan, S. Kandasamy, Eco-friendly cost-effective approach for synthesis of copper oxide nanoparticles for enhanced photocatalytic performance, Optik, 202 (2020) 163507. https://doi.org/10.1016/j.ijleo.2019.163507

[108] A.A. Olajire, A.A. Mohammed, Green synthesis of nickel oxide nanoparticles and studies of their photocatalytic activity in degradation of polyethylene films, Adv. Powder Technol., 31 (2020) 211-218. https://doi.org/10.1016/j.apt.2019.10.012

[109] R. Li, Z. Chen, N. Ren, Y. Wang, Y. Wang, F. Yu, Biosynthesis of silver oxide nanoparticles and their photocatalytic and antimicrobial activity evaluation for wound healing applications in nursing care, J. Photoch. Photobio. B, 199 (2019) 111593. https://doi.org/10.1016/j.jphotobiol.2019.111593

[110] B.N. Rashmi, S.F. Harlapur, B. Avinash, C.R. Ravikumar, H.P. Nagaswarupa, M.R.A. Kumar, K. Gurushantha, M.S. Santosh, Facile green synthesis of silver oxide nanoparticles and their electrochemical, photocatalytic and biological studies, Inorg. Chem. Commun., 111 (2020) 107580. https://doi.org/10.1016/j.inoche.2019.107580

[111] J. Ebrahimian, M. Mohsennia, M. Khayatkashani, Photocatalytic-degradation of organic dye and removal of heavy metal ions using synthesized SnO_2 nanoparticles by *Vitex agnus-castus* fruit via a green route, Mater. Lett., 263 (2020) 127255. https://doi.org/10.1016/j.matlet.2019.127255

[112] R. Magudieshwaran, J. Ishii, K.C.N. Raja, C. Terashima, R. Venkatachalam, A. Fujishima, S. Pitchaimuthu, Green and chemical synthesized CeO_2 nanoparticles for photocatalytic indoor air pollutant degradation, Mater. Lett., 239 (2019) 40-44. https://doi.org/10.1016/j.matlet.2018.11.172

[113] A.A. Ezhilarasi, J.J. Vijaya, K. Kaviyarasu, L.J. Kennedy, R.J. Ramalingam, H.A.A. Lohedan, Green synthesis of NiO nanoparticles using *Aegle marmelos* leaf

extract for the evaluation of in-vitro cytotoxicity, antibacterial and photocatalytic properties, J. Photoch. Photobio. B, 180 (2018) 39-50. https://doi.org/10.1016/j.jphotobiol.2018.01.023

[114] B. Fatima, S.I. Siddiqui, R. Ahmed, S.A. Chaudhry, Green synthesis of f-CdWO$_4$ for photocatalytic degradation and adsorptive removal of Bismarck Brown R dye from water, Water Resour. Ind., 22 (2019) 100119. https://doi.org/10.1016/j.wri.2019.100119

[115] M. Honarmand, M. Golmohammadi, A. Naeimi, Biosynthesis of tin oxide (SnO$_2$) nanoparticles using jujube fruit for photocatalytic degradation of organic dyes, Adv. Powder Technol., 30 (2019) 1551-1557. https://doi.org/10.1016/j.apt.2019.04.033

[116] A.K.H. Bashir, L.C. Razanamahandry, A.C. Nwanya, K. Kaviyarasu, W. Saban, H.E.A. Mohamed, S.K.O. Ntwampe, F.I. Ezema, M. Maaza, Biosynthesis of NiO nanoparticles for photodegradation of free cyanide solutions under ultraviolet light, J. Phys. Chem. Solids, 134 (2019) 133-140. https://doi.org/10.1016/j.jpcs.2019.05.048

[117] I. Bibi, N. Nazar, M. Iqbal, S. Kamal, H.N. Bhatti, S. Nouren, Y. Safa, K. Jilani, M. Sultan, S. Ata, F. Rehman, M. Abbas, Green and eco-friendly synthesis of cobalt-oxide nanoparticle: Characterization and photo-catalytic activity, Adv. Powder Technol., 28 (2017) 2035-2043. https://doi.org/10.1016/j.apt.2017.05.008

[118] P. Panchal, D.R. Paul, A. Sharma, D. Hooda, R. Yadav, P. Meena, S.P. Nehra, Phytoextract mediated ZnO/MgO nanocomposites for photocatalytic and antibacterial activities, J. Photoch. Photobio. A, 385 (2019) 112049. https://doi.org/10.1016/j.jphotochem.2019.112049

[119] S. Sadhukhan, A. Bhattacharyya, D. Rana, T.K. Ghosh, J.T. Orasugh, S. Khatua, K. Acharya, D. Chattopadhyay, Synthesis of RGO/NiO nanocomposites adopting a green approach and its photocatalytic and antibacterial properties, Mater. Chem. Phys., 247 (2020) 122906. https://doi.org/10.1016/j.matchemphys.2020.122906

[120] I. Fatimah, I. Sahroni, O. Muraza, R. Doong, One-pot biosynthesis of SnO$_2$ quantum dots mediated by *Clitoria ternatea* flower extract for photocatalytic degradation of rhodamine B, J. Environ. Chem. Eng., 8 (2020) 103879. https://doi.org/10.1016/j.jece.2020.103879

[121] K. Rajendaran, R. Muthuramalingam, S. Ayyadurai, Green synthesis of Ag-Mo/CuO nanoparticles using *Azadirachta indica* leaf extracts to study its solar photocatalytic and antimicrobial activities, Mat. Sci. Semicon. Proc., 91 (2019) 230-238. https://doi.org/10.1016/j.mssp.2018.11.021

[122] S. Sohrabnezhad, A. Seifi, The green synthesis of Ag/ZnO in montmorillonite with enhanced photocatalytic activity, Appl. Surf. Sci., 386 (2016) 33-40. https://doi.org/10.1016/j.apsusc.2016.05.102

[123] R. Subbiah, S. Muthukumaran, V. Raja, Biosynthesis, structural, photoluminescence and photocatalytic performance of Mn/Mg dual doped ZnO nanostructures using *Ocimum tenuiflorum* leaf extract, Optik, 208 (2020) 164556. https://doi.org/10.1016/j.ijleo.2020.164556

[124] K. Pakzad, H. Alinezhad, M. Nasrollahzadeh, Green synthesis of Ni@Fe_3O_4 and CuO nanoparticles using *Euphorbia maculata* extract as photocatalysts for the degradation of organic pollutants under UV-irradiation, Ceram. Int., 45 (2019) 17173-17182. https://doi.org/10.1016/j.ceramint.2019.05.272

[125] R. Khan, M.H. Fulekar, Biosynthesis of titanium dioxide nanoparticles using *Bacillus amyloliquefaciens* culture and enhancement of its photocatalytic activity for the degradation of a sulfonated textile dye Reactive Red 31, J. Colloid Interf. Sci., 475 (2016) 184-191. https://doi.org/10.1016/j.jcis.2016.05.001

[126] R.M. Tripathi, A.S. Bhadwal, R.K. Gupta, P. Singh, A. Shrivastav, B.R. Shrivastav, ZnO nanoflowers: novel biogenic synthesis and enhanced photocatalytic activity, J. Photoch. Photobio. B, 141 (2014) 288-295. https://doi.org/10.1016/j.jphotobiol.2014.10.001

[127] K. Cao, M.M. Chen, F.Y. Chang, Y.Y. Cheng, L.J. Tian, F. Li, G.Z. Deng, C. Wu, The biosynthesis of cadmium selenide quantum dots by *Rhodotorula mucilaginosa* PA-1 for photocatalysis, Biochem. Eng. J., 156 (2020) 107497. https://doi.org/10.1016/j.bej.2020.107497

Bioinspired Nanomaterials for Energy and Environmental Applications   Materials Research Forum LLC
Materials Research Foundations **121** (2022) 117-140   https://doi.org/10.21741/9781644901830-4

Chapter 4

Bioinspired Nanostructured Materials for Energy-Related Electrocatalysis

M. Rajkumar[1], C. Pandiyarajan[2] and P. Rameshkumar[1]*

[1]Department of Chemistry, School of Advanced Sciences, Kalasalingam Academy of Research and Education, Krishnankoil-626126, Tamil Nadu, India

[2]School of Chemistry, Madurai Kamaraj University, Madurai - 625 021, Tamil Nadu, India

*rameshkumar.p@klu.ac.in

Abstract

Conventional synthetic methods are facing great challenges to prepare functional nanostructures with fine design, tunable property, high efficiency and good sustainability. In recent decades, bioinspired synthesis has been extensively applied for the synthesis of nanomaterials with fascinating properties. Modifying the electrodes with bioinspired nanomaterials is of great interest because of their unique advantages and outperforming characteristics. In this chapter, the recent progresses on the bio-inspired synthesis of nanomaterials and their applications in energy-related electrocatalysis are focussed. The general mechanisms of key electrocatalytic processes such as oxygen evolution reaction (OER), hydrogen evolution reaction (HER), oxygen reduction reaction (ORR), methanol oxidation and formic acid oxidation reactions are discussed. Importantly, the characterization of bio-inspired nanomaterials and their enhanced energy-relevant electrocatalytic properties in terms of onset potential, peak current density and durability are elaborately reviewed. The chapter is concluded with the advantages and limitations of bioinspired methodology and the possible solutions to improve the electrocatalytic performance in the future.

Keywords

Bioinspired Nanomaterials, Oxygen Evolution Reaction, Hydrogen Evolution Reaction, Oxygen Reduction Reaction, Small Organic Molecule Oxidation Reaction

Contents

**Bioinspired Nanostructured Materials for
Energy-Related Electrocatalysis**..117

1.   **Introduction**..118

2.   **Energy-related electrocatalytic processes** 119

3.   **Synthesis and characterization of key bioinspired materials**............121

4.   **Applications of bioinspired materials in energy-related
electrocatalytic processes**...125

    4.1.   Bioinspired nanomaterials as oxygen reduction reaction
        electrocatalyst 125

    4.2.   Bioinspired nanomaterials for hydrogen evolution and
        oxygen evolution reaction 127

    4.3.   Bioinspired nanomaterials for small organic molecule
        oxidation reaction 131

Conclusions and future perspectives...134

References..134

1.   Introduction

The development of renewable and sustainable energy sources is an absolute requirement in order to face contemporary energy challenges. Electrochemical energy storage and conversion systems have received greater attention because of their sustainable features such as high efficiency, renewability, low-cost and long durability [1, 2]. However, the development of electrode materials featured by environmental sustainability, smart structure and intelligent function is a challenging task. Traditional synthetic methodologies are confronted with great challenges to fabricate delicate nanostructured electrode materials [3]. Therefore, the idea of bioinspired synthesis to overcome these barriers is undoubtedly promising and feasible due to the vast diversity of nature. Along with microorganisms and plants, viruses [4], proteins [5] and DNA [6] have also become potentially useful candidates for bio-inspired synthesis of nanostructured materials.

The hydrogenase metalloenzymes present in microorganisms and algae are capable of reversibly converting protons into molecular hydrogen [7]. The active sites of these enzymes are made up of Fe/S or Fe/Ni/S core clusters, and molecular hydrogen can be evolved at turnover frequencies as high as 9000 moles of H_2 per mole of hydrogenase per second in water at pH 7 and 30 °C [8]. Hydrogenase biomimetic synthetic molecular electrocatalysts show poor to moderate activity toward molecular hydrogen evolution or uptake reactions in solution or grafted onto an electrode even though it has attractive properties such as their oxygen stability and solubility in different media [9]. Natural DNA has emerged as an attractive biomacromolecule for electronic, optical, and biomaterials [10]. It is abundant, renewable, and biodegradable and possesses many unique properties. DNA's rich chemical functionality allows it to interact with a variety of nanomaterials of interest. The aromatic nucleobases in DNA can interact through $\pi-\pi$ stacking with graphene basal surface [11]. The transition metal cations can be chelated with the aromatic nucleobases in DNA *via* dative bonding. As a highly charged polyelectrolyte, the negatively charged phosphate groups on the DNA backbone can endow DNA-modified materials with high dispersibility in aqueous solutions. Therefore, these properties of natural DNA make it the ideal mediator to build graphene-based metal catalyst. Melanin biomass and biomimetic melanin-like polymer could be good precursors for highly N-doped carbons that potentially present enhanced CO_2 uptake and efficient ORR activity [12]. Melanin is a well-known biopolymer that is broadly distributed in most of living organisms, which has shown characteristic features such as protection of human and animal skin from UV light, free radical quenching, metal ion chelation, antibiotic, thermoregulation, and so on [13]. The bio-inspired hierarchical "grape cluster" superstructure has provided an effective integration of one-dimensional carbon nanofibers (CNF) with isolated carbonaceous nanoparticles into three-dimensional (3D) conductive frameworks for efficient electron and mass transfer in ORR [14]. Enzymes, particularly metalloenzymes, demonstrate high activity towards many energy transformation reactions in biological systems, including ORR [15]. The synthesis of novel electrode materials for electrochemical energy storage and conversion can be deeply connected to the inspirations of nature, which provides more flexible and intelligent options.

In this chapter, the recent progresses in the highly promising field of bioinspired synthesis of nanomaterials for energy-related electrocatalytic processes are specifically focussed. Firstly, basic principles of key energy-related electrocatalytic processes, and synthesis and characterization of some important bioinspired materials are discussed.   Then the applications of bioinspired materials such as HER, OER, ORR, methanol oxidation and formic acid oxidation are elaborately discussed.

2.   Energy-related electrocatalytic processes

Electrocatalysis research plays a vital role in the interconversion of electrical and chemical energy. Frequently encountered energy-related electrocatalytic processes are oxygen reduction reaction, oxygen evolution reaction, hydrogen evolution reaction, methanol oxidation, ethanol oxidation, formic acid oxidation and hydrazine oxidation. Oxygen reduction reaction, in the fuel cell mode, occurs for the energy conversion [16]. The ORR usually involves multistep electron-transfer processes. It is more favourable in alkaline medium than in acidic environments under ambient temperature due to improved kinetics and lower overpotential. Oxygen can be directly reduced to water by accepting four electrons per O_2 molecule. On the other hand, it can be reduced into H_2O_2 by accepting only two electrons [17]. Therefore, oxygen reduction may involve a four-electron ($4e^-$) or a two-electron ($2e^-$) transfer reaction on surfaces of metals. The overall electron transfer numbers per oxygen molecule involved in the typical ORR process are calculated from the slopes of the Koutecky-Levich plots [18]. The mechanism of electron transfer process may vary with the catalytic nanomaterials applied in ORR. The ORR behaves as a structure-sensitive reaction even for the same catalyst. The O_2 adsorption configurations and the interaction of O_2 with materials surface can affect the ORR pathways.

Hydrogen evolution reaction and oxygen evolution reaction are conducted for energy storage applications in the electrolyzer mode. In general, the overall reaction of water splitting consists of both reduction (HER) and oxidation (OER) half reactions. Electrocatalytic HER on catalytic sites involves two different processes to form H_2 by the reduction of protons [19]. In the first step, both the processes primarily involve the reduction of proton into hydride species on catalytic centers. In the second step, two hydride species can react to form H−H in one mechanism. In the other mechanism, one hydride reacts with one proton−electron couple to form one hydrogen molecule. The catalytic materials are used in order to minimize the overpotential of H_2 evolution reaction. Platinum is considered as the best known catalyst for HER that requires very less overpotentials even at high reaction rates in acidic solutions. In OER, molecular oxygen is produced via several proton/electron coupled reactions [20]. In acidic and neutral conditions, two water molecules are oxidized into four protons (H^+) and O_2, while hydroxyl groups (OH^-) are oxidized and transformed into H_2O and O_2 in basic environments. The redox reactions involved in water splitting are schematically shown in Figure 1.

The direct methanol fuel cells (DMFCs) are one of the most promising energy sources for portable devices and transportation applications. The appealing features of a DMFCs include high efficiency, high energy density, low operating temperature, environmentally benign quality, and the ability to generate power without distributed pollution from by-products of hydrocarbon combustion [21]. The methanol electro-oxidation reaction is a slow process and it involves the transfer of six electrons to the electrode for complete oxidation to carbon dioxide. Various reaction intermediates may be formed during methanol oxidation. Some of these (CO-like) species are irreversibly adsorbed on the surface of the electrocatalyst which has the effect of significantly reducing the fuel consumption efficiency and the power density of the fuel cell [22]. Direct formic acid fuel cells (DFAFCs) are also considered as one of the promising energy conversion devices due to their high electromotive force, safety of fuels, limited fuel crossover, and high practical power density at low temperature [23]. Pt is considered as an active element for the adsorption of formic acid and it is frequently used as an electrocatalyst because of its high acid tolerance. It is well known from the literature reports that there are three different path ways for the electrochemical oxidation of formic acid on Pt surface: (i) formation of CO_2 with the elimination of two hydrogen atoms, (ii) formation of CO intermediate and subsequent CO oxidation, and (iii) formation of formate ion with the removal of hydrogen atom and its oxidation to CO_2 [24]. Hydrazine as a fuel for alkaline fuel cells has been studied for many decades. It has received a great attention as a promising fuel due to the reasons such as non-emission of CO_2, zero production of poisonous species and hogh power density [25].

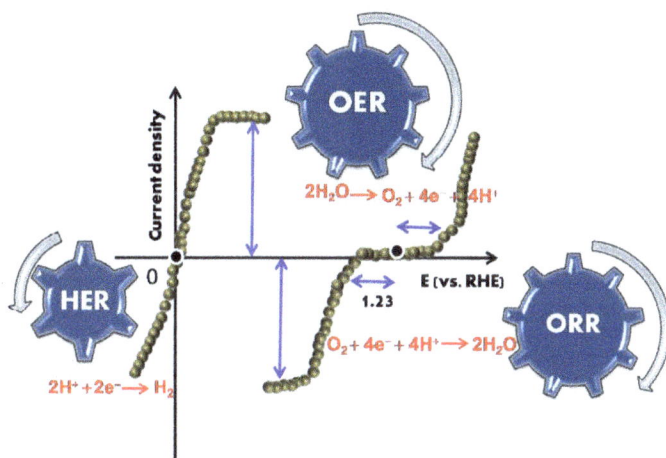

Figure 1. Schematic representation of redox reactions involved in water splitting. Reproduced permission from Ref. No. [17].

3.   Synthesis and characterization of key bioinspired materials

Huang et al. designed and synthesized three-dimensional graphene aerogel (GA) supported FeN_5 composite with five coordinated Fe-N bond [26]. The bifunctional linker of 4-AP strongly coordinate with FePc molecules through strong dative bond and the formed FePc/AP complexes attaches with Graphene aerogel (GA) via a reduction induced solution self-assembly process. The unique structure of GA not only increased the surface area but also facilitated the decoration of FePc molecules on both sides of the graphene sheets. The FT-IR spectra results indicated the anchoring of FePc molecules on graphene surface through coordination interaction of 4-AP linker at the axial position. Cao et al synthesized the composite of FePc and single-walled CNTs (FePc–Py–CNTs) from covalent functionalization of single-walled CNTs (Py–CNTs) [27]. Thermal gravimetric analysis of FePc–Py–CNTs indicated the stoichiometric ratio of Py/C) is 1/48. The higher D-band intensity of FePc–Py–CNTs composite in raman spectra is another indication for the covalent functionalization of CNTs. X-ray photoelectron spectroscopy (XPS) illustrates the more positive nature of iron atom due to the presence of extra pyridine ligand in axial position. The 3D flower-like g-C_3N_4@carbon (CN@C) nanosheets was synthesized through a novel bio-inspired approach [28]. The sucrose molecule was used as the inducer to control the morphology of CN@C nanosheets. SEM image reveals the flower-like architecture is porous, fluffy and highly interconnected, which is formed from uniformly distributed nanosheets (Figure 2). The thickness of nanosheets are found to be 15 ± 2.2 nm. In FT-IR spectrum of CN, three characteristic absorption bands are observed at 3000–3500, 1200–1650 and 810 cm^{-1} which correspond to the stretching mode of N–H bond, stretching modes of heterocycles and breathing mode of s-triazine, respectively. XPS results confirm the existence of the heptazine heterocyclic ring (C_6N_7) unit, which is the primary building block of g-C_3N_4. While, the peak at 531.5 eV in O1s spectrum may be attributed to the –COOH bond from carbon layers. The specific surface area was calculated to be 208.7^{2+}g^{-1} via the Brunauer-Emmett-Teller method. PL results indicate that the recombination of electron-hole pairs is dramatically prohibited in the 3D CN@C microflowers.

Giovanni et al. prepared air-stable pyrrhotite-type FeS nanoparticles by solvothermal method [29]. The TEM image shows hexagonally shaped nanoparticles with size range from 50 to 500 nm which is in close agreement with the particle size calculated from XRD. From X-ray diffraction, it was established that the sample is well-crystalline and the broadened lines should result from local change in the environment of the Fe nuclei, which is observed in the Mossbauer spectra (Figure 3). Y. Yan et al. synthesized hierarchical FeP nanoarray films composed of FeP nanopetals synthesized through a bio-inspired hydrothermal route followed by phosphorization [30]. The SEM image of the α-FeO(OH) nanopetals, reveals that the petals have a very smooth surface, along with a thickness of 5–10 nm and a lateral size of 200–300 nm. The concentration of glycerol influenced the morphology of the product which is monitored by the time dependent experiments. The AuPt nanodendrites (NDs) were synthesized by a one-pot wet-chemical bio-inspired strategy at room temperature [31]. In this synthesis, Folic acid (FA) used as the weak stabilizer, structure director, and capping agent in the formation of AuPt NDs. High-resolution TEM image confirms the nanodentrite structure of AuPt and also verifies the alloy nature of the catalyst, which is in good agreement with the result obtained from the SAED pattern (Figure 4). XPS demonstrates that the Pt^{2+} and Au^{3+} completely reduced to Pt^0 and Au^0, respectively. Qu et al. synthesized different Graphene-Based Pd Catalysts using DNA and PVP as stabilizers to enhance the dispersibility of the reduced graphene and the growth of Pd nanoparticles (NPs) at the surface of graphene (Figure 5) [32]. The disappearance of the absorption of H_2PdCl_4 at 425 nm in UV-vis spectroscopy indicated Pd^{2+} has been reduced to Pd NPs. TEM images of GO and DNA-G and PVP-G composites are well-dispersed ultrathin flat sheets ranging from 200 nm to more than 1 μm in size and have occasional folds, crinkles, and rolled edges, indicating that both DNA and PVP can stabilize the reduced graphene ascribing to the π−π interaction and hydrophobic interaction, respectively. Both DNA-G-Pd and PVP-G-Pd have a rougher surface than the corresponding graphene sheets, indicating Pd NPs are facilely adsorbed on the surface of graphene nanosheets which is illustrated in AFM. The XRD pattern of DNA-G-pd reveals the crystalline nature of the composite. From Raman spectra, the value of I_D/I_G (1.09) for DNA-G-Pd is greater than that of DNA-G, which indicates the interactions occurred between Pd nanoparticles and DNA-G. Nitrogen 1s and phosphorus 2p XPS data confirmed that the DNA molecules were adsorbed onto the graphene sheets and peaks at 335.1 and 340.0 eV indicates the reduction of Pd (II) to Pd (0). The DNA-G and PVP-G composites contain about 15 wt % DNA and 24 wt % PVP, which is observed in TGA.

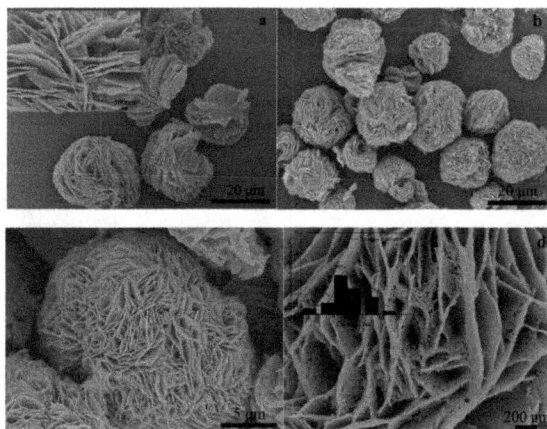

Figure 2. SEM images of (a) 3D CAM@C (the inset is its magnification) and (b–d) CN@C microflowers with different magnifications. The inset in Fig. 2d is the thickness distribution of 2D CN@C nanosheets. Reproduced permission from Ref. No. [28]

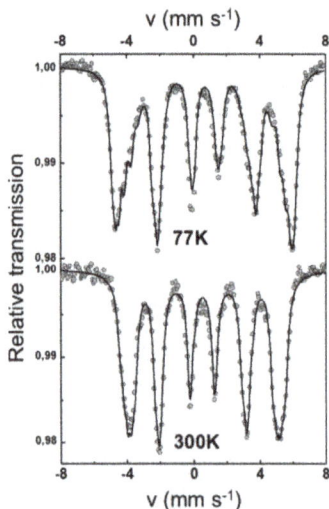

Figure 3. Mossbauer spectra recorded at 300 and 77 K. Reproduced permission from Ref. No.29

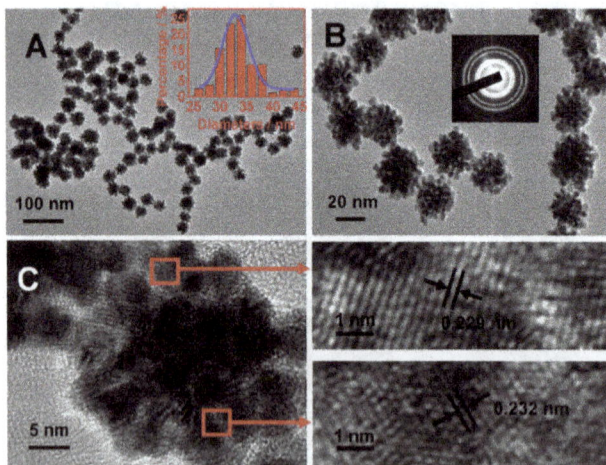

Figure 4. Low- (A), medium- (B), and high-resolution (C) TEM images of Au₆₅Pt₃₅ NDs. Inset in Figure 2B shows the SAED pattern. Reproduced permission from Ref. No. [31].

Bioinspired Nanomaterials for Energy and Environmental Applications   Materials Research Forum LLC
Materials Research Foundations **121** (2022) 117-140   https://doi.org/10.21741/9781644901830-4

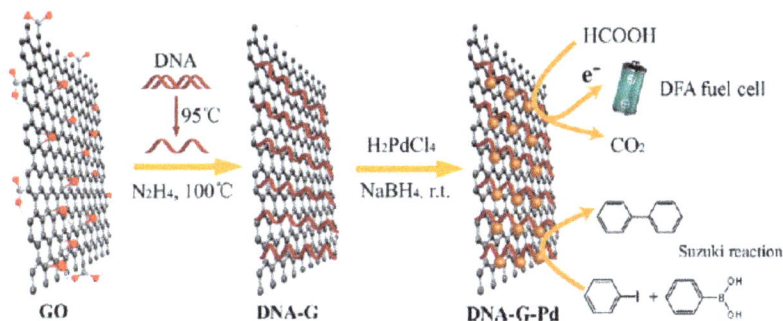

Figure 5. Schematic Illustration of the Procedure to Design Calf Thymus DNA-Modified Graphene/Pd Hybrid and the Catalysis Application in Direct Formic Acid (DFA) Fuel Cell and Suzuki Reaction. Reproduced permission from Ref. No.[32].

4.   Applications of bioinspired materials in energy-related electrocatalytic processes

4.1.   Bioinspired nanomaterials as oxygen reduction reaction electrocatalyst

The electrocatalytic properties of iron and manganese coordinated metal organic framework on Au (111) surface has been explored towards oxygen reduction in alkaline media [33]. The trimesic acid (TMA) and bis-pyridyl-bispyrimidine (BPA) networks coordinating to single iron and manganese atoms on Au (111) effectively catalysed the oxygen reduction. The shoulder observed at -0.2 V shows an onset potential shifted 0.05 V to lower overpotential and has tripled current density (0.75 mA cm^{-2}) compared with bare Au (111). The other striking feature is the presence of a second peak at about -0.80 V with a large current density of 2.5 and 3 mA cm^{-2} for TMA–Fe and PBP–Fe, respectively. But only small differences observed between TMA–Fe and PBP–Fe polarization curves. The polarization curve for TMA–Mn in O$_2$-saturated 0.1 M NaOH solution shows distinct behaviour in comparison to bare Au (111) and TMA-Fe, PBP-Fe networks. The polarization curve is shifted to a lower potential of -0.15 V with an onset potential lowered by 0.1 V and shows a high current density of 1 mA cm^{-2}. The networks containing Fe atoms follows (2+2) e$^-$ pathway for the reduction of O$_2$, while the coordinated Mn atoms follows direct 4 e$^-$ route to convert O$_2$ to H$_2$O. The chemical activity of the metal centres is determined by both the nature of the metal ion and its coordination shell. The electrocatalytic oxygen reduction in alkaline media was investigated using the FePc–Py–CNTs composite by Cao *et al.* [27]. The FePc–Py–CNTs exhibited higher ORR activity than that of FePc–CNTs and Py–CNTs. The half-wave potential (E$_{1/2}$) for the

FePc–Py–CNTs composite was more positive by~35mV than the commercial Pt/C catalyst. The synthesized FePc–Py–CNTs also show higher ORR activity in acid medium. To study the mechanism of ORR, KCN was introduced in the reaction mixture which significantly inhibits the activity of synthesized composite. This clearly demonstrates that the Fe ion plays active role in the catalyst. Further, after the removal of Cyanide ion through rinsing restores the activity of the catalyst. The number of electrons involved in ORR for FePc–Py–CNTs was found to be~4.05 from Koutecky–Levich's equation suggesting the direct four electron pathway. The FePc–Py–CNTs composite retained~92% of the initial current density whereas commercial Pt/C catalyst retains only~30% of current density after 1000 cycles in the durability test. The practical application of this catalyst was also validated in Zn-air batteries, where the catalyst acts as the best cathode material. Spin-polarized density functional theory (DFT)-based calculations discussed the electronic structure change of Fe centre in aromatic macrocycle and which is responsible for the excellent ORR activity. To assess the ORR catalytic activity of pyridine modified Graphene anchored iron phthalocyanine molecules (FePc/AP-GA), cyclic voltammetry (CV) was performed in N_2- and O_2-saturated in 0.1 M KOH solution [26]. The composite showed a half-wave potential of -0.035 V (vs. Hg/HgO) in alkaline electrolyte, which is more effective in ORR activity than benchmarked Pt/C and most of the pyrolyzed or nonpyrolyzed nonprecious metal catalysts. The FePc/AP-GA composite also exhibits high kinetic current density of 20.01 mA cm^{-2} at –0.1 V, good durability, and high tolerance to methanol poisoning effects. The geometric and electronic structure of iron atom in FePc/AP-GA may facilitate the adsorption of O_2 and intermediates, which enhance ORR activity and durability. Linear Sweep Voltammetry was recorded for carbon nanofibers with grape-like N-doped hollow carbon particles (CNF@NC) by rotating ring-disk electrode in oxygen saturated KOH solution (Figure 6) [14]. It is inferred that the CNF@NC composite with optimum pore size shows high limited current density of 7.7 mA cm^{-2} at 0.589 V, which is much higher than that of controlled samples. The metal-free CNF@NC catalyst demonstrates superior catalytic activity, follows four-electron transfer mechanism, good methanol tolerance and having long-term stability. The effective oxygen absorption ascribed to the efficient 3D conductive pathways with the nitrogen-induced charge delocalization and high hydrophilicity of the highly porous NC particles [34, 35].

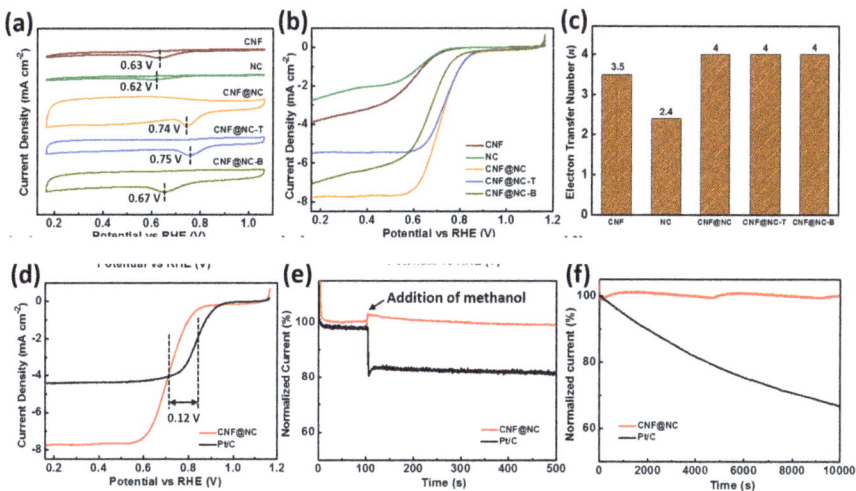

Figure 6. (a) CV curves of different sample-modified electrodes in O_2-saturated 0.1 M KOH solution with a sweep rate of 5 mV s^{-1}. (b) LSV curves of different sample-modified electrodes at the rotation rate of 1600 rpm in O_2-saturated 0.1 M KOH solution. (c) The calculated electron transfer numbers of different samples. (d) The comparison of the $E_{1/2}$ potentials for the CNF@NC and commercial Pt/C electrode. (e) Chronoamperometric responses of CNF@NC and Pt/C at 0.589 V vs. RHE O_2-saturated 0.1 M KOH solution. The arrow indicates the addition of methanol. (f) Chronoamperometric responses of CNF@NC and Pt/C electrode at 0.589 V vs. RHE in O_2-saturated 0.1 M KOH solution. Reproduced permission from Ref. No. [14].

4.2. Bioinspired nanomaterials for hydrogen evolution and oxygen evolution reaction

The electrocatalytic properties of bioinspired air-stable pyrrhotite-type FeS nanoparticles dispersed in Nafion films was studied for Hydrogen evolution from neutral water [29]. FeS nanoparticles exhibited a sharp rise in current from ca. − 0.8 vs NHE in 0.1 mol/L potassium phosphate (pH 7.0) and small bubbles evolved from the surface of the coated electrode due to the evolution of Hydrogen. In the reverse scan, two oxidation peaks at − 0.50 and − 0.42 vs NHE, may be due to oxidation of hydride or dihydrogen species trapped at the surface of the catalytic film. The coated electrodes were stable for 6 months without proper care. Galvanostatic experiments were conducted to check the stability of modified electrode at a current density of J = 0.7 mA cm^{-2} (I = 50 µA) at pH 7.0. During the first 24 h, decrease in overpotential was observed and the activity of the film toward H_2 production was enhanced. The small spikes observed may be due to the formation of small H_2 bubbles at the surface of the coated electrode. On varying the pH from 6 to 8, higher catalytic activity was achieved at lower pH and strong decrease in the current density was observed in basic conditions. The change in pH did not produce any irreversible alterations of the film and as the initial current density was attained while reverting to the initial pH 7.0. The durability and robustness of the FeS nanoparticle-coated electrode was assessed by performing long-duration controlled-potential electrolysis (CPE). Quantitative Faradaic yield of ≥ 0.99 was attained for the modified electrode in the evolution of Hydrogen. Tran N. Huan *et al* investigated the catalytic activity of the nickel bis(diphosphine) functionalized MWCNT electrode ($[Ni(P^{Ph}_2N^{R1}_2)_2]^{2+}$ and $[Ni(P^{Cy}_2N^{R1}_2)_2]^{2+}$) has been carried out in 0.5 M H_2SO_4 aqueous solution [36]. The current density obtained for hydrogen production at -300 mV vs SHE is slightly higher for NiP^{Ph}_2 than for NiP^{Cy}_2 material. The catalysts with phenyl groups at the phosphorous are more pertinent to hydrogen production, while those with cyclohexyl groups are more prone to hydrogen oxidation and the findings are in good agreement with reported literatures [37, 38]. Increased current density is directly related to the increased amount of grafted NiP^{Cy}_2 catalyst on MWCNTs drop casting electrode when compared to filtration method and the calculated value found to be 5.0 ($\pm$0.5) $\times 10^{-9}$ mol.cm^{-2}. This clearly shows that the method of constructing the electrode material influences the catalyst efficiency. The introduction of carbon fibers into the GDL electrode provides ultra-high surface area and more permeable to hydrogen gas. The best performances were obtained at the optimum quantities of 0.267 mg MWCNTs and 0.033 mg carbon microfiber per square centimeter of GDL. The increased surface area of the porous structure increases the amount of NiP^{Cy}_2 loaded on the electrode and a fivefold increase compared with the drop-cast MWCNT/GDL electrode. The MWCNT+carbon microfiber/GDL electrode showed catalytic activity of 16 mA cm^{-2} at -300 mV vs SHE for hydrogen production and diffusion of the hydrogen gas through the porous structure have been achieved at the catalyst active site. The catalytic activity of the electrode was investigated at the elevated temperature. The polarizations of the electrode were recorded at regular intervals of 10 °C on varying the temperature from 25 °C to 85 °C. At elevated temperature, the electrode showed fivefold increase of current in the production of Hydrogen. The durability of the MWCNT+carbon microfiber electrode for hydrogen production was conducted at -100 mV vs SHE and withstands over 7 hours of electrolysis.

Y. Yan *et al.* obtained the polarization curves for nanostructured FeP film electrodes for Hydrogen evolution reaction both in acid and base media [30]. In 0.5 M H_2SO_4, a rapid cathodic response was observed for nanostructured FeP film and further scanning gave rise steep increase in current density, which exceeds that of commercial Pt/C. The onset overpotential of the FeP film electrode was calculated to be ~16 mV, and overpotential of 65 mV was only sufficient to drive a kinetic current density of 10 mA cm^{-2}. The Tafel slope for FeP film and commerical Pt/C electrode were 48.5 mV dec^{-1} and 30.0 mV dec^{-1}, respectively. The slope value indicates the mechanism involved in HER was quite different and Volmer-Heyrovsky mechanism followed in FeP film electrode. The mechanism implies bond strength of the adsorbed hydrogen is not strong enough desorbs the product and it only provide coverage of the intermediate [39]. In alkaline electrolyte, the onset overpotential of the FeP nanostructured film was found to be 16 mV, and only requires overpotential of 84 mV to achieve a geometric current density of 10 mA cm^{-2}. The FeP film can produce a catalytic current density of 100 mA cm^{-2} at a potential of -193 mV vs. RHE, which is quite higher than that of commercial Pt/C electrocatalyst. A smaller slope of 85.0 mV dec^{-1} for FeP film was observed in Tafel plots compared with Pt/C (120.5 mV dec^{-1}), further suggesting the favorable electrochemical kinetics for the HER process. Likewise, same Volmer-Heyrovsky mechanism was followed in basic media except that rapid recombination between H$^+$ cations and the abundant OH$^-$ anions takes place [39]. Thus, the HER is mainly influenced by hydrogen binding energy in both strong acids and strong bases. The FeP array film exhibited excellent long-term stability for more than 5000 cycles only with slight degradation of current density in both medias. The intial morphology of FeP array film was maintained after several stability tests. The nanopetal structure of FeP films played a major role in the performance of HER. To understand the catalytic activity of FeP array film in HER, density functional theory calculations were perfomed which revealed that the mixed P/Fe termination in the FeP film is responsible for the high catalytic activity of the nanostructured electrodes. The electrocatalytic performance of the Cobalt-Phosphonate Nanosheets (1Co-ns) modified GC electrode for HER was investigated in N$_2$-saturated tris-(hydroxymethyl)aminomethane-HNO$_3$ (tris-HNO$_3$) neutral aqueous solution (pH 7.4) [40]. It has been shown that the 1Co-ns-modified GC electrode exhibits superior HER activity with an onset potential of approximately $-$ 0.084 V against RHE (Figure 7). The optimum concentration of 1-Co-ns was found to be 2 mg/mL. For comparison, the electrocatalytic activity of 1-Co towards HER was investigated which showed an onset potential of $-$ 0.186 V and it is considerably more negative than that for 1Co-ns. The enhanced activity could be ascribed to the greater exposure of the catalytic active sites and improved charge transport in the two dimensional nanosheet catalysts. The rate limiting step during the HER process was studied via Tafel slope. Three principal steps have been followed in HER: the Volmer (discharge), Heyrovsky (electrochemical desorption), and Tafel (recombination desorption) steps with Tafel slopes of 116, 38, and 29 mV/dec, respectively. To clarify the the contribution of the Volmer$-$Heyrovsky and Volmer$-$Tafel mechanisms, deuteration in HER was conducted.

From the observed Tafel slope, it is concluded that HER is largely influenced by the deuterated effect, which suggests that electrochemical desorption is the rate-limiting step and thus the Volmer−Heyrovsky mechanism is more effective in the HER catalyzed by 1Co-ns. The 1Co-ns catalyst shows exceptional stability as indicated by the negligible decline in the overpotential and current density within 1000 potential scans in neutral pH. under both in acidic and basic conditions, the catalyst shows poor HER activity due to protonation which weakens the coordinate bond and further result in the collapse of the skeleton within the phosphonate CPs [41-43].

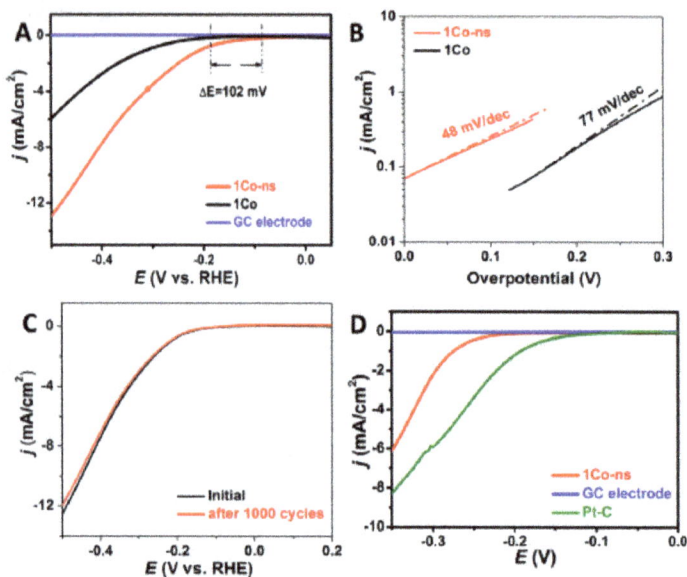

Figure 7. Electrochemical characterization of various catalysts toward HER. (A) HER polarization curves obtained on several catalysts measured in 0.1 M tris-HNO₃ (pH 7.4). (B) Tafel plots of the corresponding electrocatalysts derived from the early stages of the HER polarization curves. (C) Stability measurements by recording the polarization curves for the Co-ns catalyst before and after 1,000 potential scans in 0.1 M tris-HNO₃ (pH 7.4) with 20 mV s⁻¹. (D) HER polarization curves of various catalysts (as indicated in the figure) in artificial seawater (pH 7.02) at a scan rate of 20 mV s⁻¹. Reproduced permission from Ref. No. [40].

The OER activity of pure CN and 3D CN@C microflowers were conducted in alkaline media [28]. The linear sweep voltammograms (LSVs) indicate that 3D CN@C microflowers has higher current densities and lower OER onset potential (1.40 V for 3D CN@C microflowers vs. RHE) than pure CN (1.70 V vs. RHE). The electrocatalytic kinetics for OER was determined by Tafel plots. The Tafel slope of 3D CN@C microflowers (254.2 mV dec^{-1}) is much smaller than that of pure CN (384.1 mV dec^{-1}). The electrochemical impedance spectroscopy (EIS) was carried out to evaluate the OER reaction kinetics. The Nyquist plot of 3D CN@C microflowers reveal smaller interface charge transfer resistance of 3D CN@C microflowers which assigned to the microflower structure that can confer much faster charge transfer during the OER process. The e stability test of 3D CN@C microflowers shows only a slight change in anodic current of 17.6% is observed for 12 h. This confirms the excellent stability of 3D CN@C microflowers. The OER activities of 3D CN@C microflowers with different layers of carbon were investigated. As the sucrose concentration increases (from 0.35 to 0.70 mg mL^{-1}), the current density increases and the OER onset potential decreases. This phenomenon is primarily caused by the thickness of carbon layer with the variation of sucrose concentration. The sucrose concentration of 0.70 mg mL^{-1} renders the optimal charge transfer in 3D CN@C microflowers. The enchanced OER electrocatalytic activity of 3D CN@C microflowers could be attributed to the electron-accepting pyridinic and tertiary N species of g-C$_3$N$_4$ impart a relatively high positive charge density to the neighboring sp^2-bonded C atoms [44-46] and *in-situ* carbon layer coated on the g-C$_3$N$_4$ nanosheet and the 3D porous microflower framework [47, 48].

4.3.   Bioinspired nanomaterials for small organic molecule oxidation reaction

The electrocatalytic activity of bioinspired AuPt nanodendrites were explored towards methanol oxidation reaction (MOR) in basic media [31]. The CV curves of Au$_{82}$Pt$_{18}$ NPs, Au$_{65}$Pt$_{35}$ NDs, and Au$_{31}$Pt$_{69}$ NPs and Pt black catalysts modified electrodes for MOR were obtained in 1.0 M NaOH (Figure 8). It is clear that the different peak current densities and peak potentials observed for the catalysts. The anodic peak is attributed to the oxidation of methanol resulting in many CO-like intermediates, which would be further oxidized at higher potential owing to the formation of Au-OH [49, 50]. However, in the reverse scan the CO-like species can't be completely removed at low potential, causing the poisoning of the catalyst. Thus, the tolerance of the catalyst toward poisoning carbonaceous species was reflected in the ratio of the peak current densities in the forward to backward scans (j_f/j_b) [51]. The j_f/j_b ratios of Au$_{82}$Pt$_{18}$ NPs, Au$_{65}$Pt$_{35}$ NDs, Au$_{31}$Pt$_{69}$ NPs, and Pt black catalysts are 14.53, 11.58, 6.26, and 5.92, respectively. The higher ratio (j_f/j_b) of Au65Pt35 NDs may be due to the the higher content of Au, which can absorb more OH$_{ad}$ to eliminate the CO-like species and increases the poisoning-resistant ability. The mass activity (MA) values of $Au_{31}Pt_{69}$ NPs, $Au_{65}Pt_{35}$ NDs, and $Au_{82}Pt_{18}$ NPs and Pt black catalysts calculated to be 917, 2098, 1441 and 172 mA mg_{Pt}^{-1}, respectively. The J_f implies oxidation catalytic current and mass activity (MA) is highest for $Au_{65}Pt_{35}$ NDs ascribed due to the unique dendrite-like nanostructure and high Au content in AuPt NDs. The electrocatalytic stability of all the catalysts was recorded by chronoamperometry for MOR. Under same conditions, the current density for $Au_{65}Pt_{35}$ NDs retains 23.13% of its original value upto 1000 s, which is higher than that of $Au_{31}Pt_{69}$ NPs (7.57%), $Au_{82}Pt_{18}$ NPs (14.44%) and Pt black (4.01%). Further, after 2000 s, the current densities decreased to 16.33%, 2.90%, 5.78% and 1.45% of their initial values for Au NPs, $Au_{31}Pt_{69}$ NPs, $Au_{82}Pt_{18}$ NPs and Pt black, respectively. The above results conclude that the higher current density and slower decay speed for $Au_{65}Pt_{35}$ NDs. The better catalytic activity of $Au_{65}Pt_{35}$ NDs are attributed to their unique dendrite-like structure and enlarged ECSA, as well as the synergetic effects of Au and Pt. Similar results were obtained for ethanol oxidation reaction.

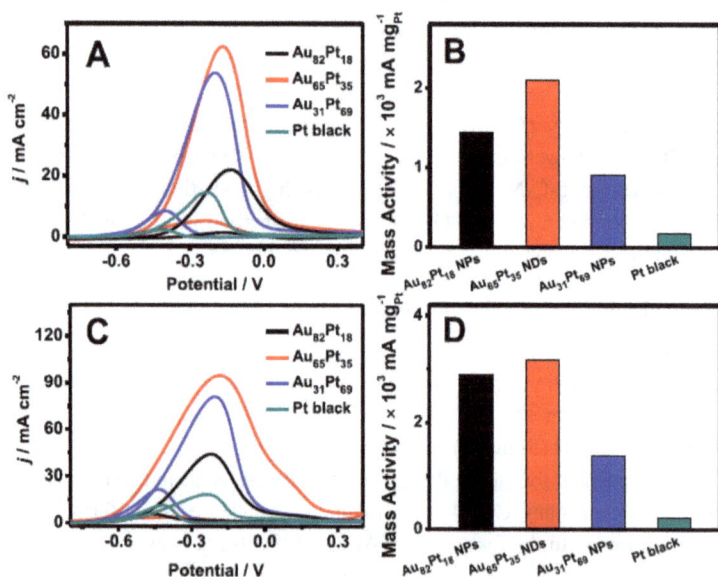

Figure 8. CV curves and the corresponding mass activities of $Au_{65}Pt_{35}$ NDs, $Au_{82}Pt_{18}$ NPs, $Au_{31}Pt_{69}$ NPs, and Pt black catalysts modified electrodes in 1.0 M KOH + 0.5 M methanol (A and B) and 1.0 M KOH + 0.5 M ethanol (C and D) at the scan rate of 50 mV s^{-1}, respectively. Reproduced permission from Ref. No. [31].

Qu et al. assessed the formic acid electro-oxidation of DNA-Modified Graphene/Pd Nanoparticles in acidic media [32]. The CVs of different Pd catalyst-coated electrodes were recorded in N_2 saturated aqueous solution of 0.5 M H_2SO_4 at a scan rate of 20 mV s^{-1} to estimate the the electrochemical active surface areas (ECSAs). The ECSAs for DNA-G-Pd, PVP-G-Pd, Pd/C, rGO-Pd, and GO-Pd catalysts were estimated to be 82.2, 54.3, 44.5, 22.5, and 1.38 cm^2 mg^{-1}, respectively. The largest ECSAs of the DNA-GPd catalyst may be due to the smaller size and good dispersibility of Pd nanoparticles on the DNA-templated graphene. The CVs of different catalyst-coated electrodes were in 0.5 M HCOOH and 0.5 M H_2SO_4 solution. Due to their poor conductance and dispersibility, both GO-Pd and rGO-Pd, show poor anodic peak current. On the other hand, the main oxidation peak potential of HCOOH for DNA-G-Pd catalyst is located at 0.25 V, which is 0.05 V more negative than that of Pd/C catalyst (0.3 V). For comparison purpose, the electrocatalytic activity of PVP-G-Pd catalyst was recorded which shows that the main oxidation peak potential is more positive and shifts to 0.65 V quite similar to Pt for HCOOH electro-oxidation [52, 53]. The mass-normalized anodic peak current of DNA-G-Pd at 0.25 V is 228.1 mA mg^{-1}, which is 2.5-fold and~3.5-fold higher than that of PVP-G-Pd (92.2 mA mg^{-1}) and Pd/C (64.8 mA mg^{-1}), respectively. The corresponding oxidation potentials of HCOOH on the DNA-G-Pd catalyst are proven to be lower than commercial Pd/C and PVP-G-Pd. Chronoamperometric measurements were carried out using different catalysts at 0.25 V in 0.5 M H_2SO_4 containing 0.5 M HCOOH to evaluate the long-term activity and durability of the catalysts. After 1000 s, the current for Pd/C catalyst decreased to 2% of the initial value, whereas for PVP-G-Pd and DNA-G-Pd catalysts, the currents still kept 11% and 15% of the initial, respectively. After 2500 s, the current for DNA-G-Pd catalyst was 15.8 mA mg^{-1}, which is 18.6-fold higher than PVP-G-Pd catalyst. After 4000 s, the current still remained at 8.0 mA mg^{-1} for DNA-G-Pd catalyst. The improved results indicates DNA act as mediator and provides heterogeneous nucleation and anchor sites, which facilitate the formation of the uniform and monodispersive Pd NPs in well-defined binding sites along DNA lattices. This brings highly electrochemical active sites on the electrode surface important for formic acid oxidation. Secondly, it can interact strongly with graphene sheets through π–π stacking and can also chelate Pd via dative bonding and increase the catalyst stability. Third, the presence of nucleobases of DNA, such as guanine and adenine deplete oxygen in HCOOH electrooxidation solution and favour the conversion of Pd to PdO and enhance the efficiency of the formic acid electrooxidation. [54, 55].

Conclusions and future perspectives

By the inspiration of fast growing nanotechnology, researchers are currently developing functional nanomaterials with environment friendly synthetic routes. This chapter summarizes the recent advancement in the field of bioinspired synthesis and their energy-relevant electrocatalytic applications. The favorable characteristics of nanomaterials synthesized using naturally available biomaterials or commercial biomolecules made them as good alternatives to traditional materials. These bioinspired materials as electrode materials showed outstanding performances in energy-relevant electrocatalytic processes. Even though the bio-resource derived nanomaterials or artificial biomolecules offers a variety of advantages they simultaneously set huge barriers for building general relationship between the synthesis and properties. Achieving a reproducibility in bioinspired synthesis is the biggest challenge in the exploration of these materials for applications. It is very difficult to predict the performance through computational methods due to the complexity in biostructures. Fortunately, continuous progress and in-depth understanding of the bioinspired synthesis and the bio-derived electrode materials made by the researchers show the importance of this research area in electrochemical energy-related catalytic applications.

References

[1]   Y. Sun, N. Liu, Y. Cui, Promises and challenges of nanomaterials for lithium-based rechargeable batteries, Nat. Energy 1 (2016) 16071. https://doi.org/10.1038/nenergy.2016.71

[2]   G. Chen, L. Yan, H. Luo, S. Guo, Nanoscale engineering of heterostructured anode materials for boosting lithium-ion storage, Adv. Mater 28 (2016) 7580–7602. https://doi.org/10.1002/adma.201600164

[3]   S. Chu, A. Majumdar, Opportunities and challenges for a sustainable energy future, Nature 488 (2012) 294–303. https://doi.org/10.1038/nature11475

[4]   E. Dujardin, C. Peet, G. Stubbs, J. N. Culver and S. Mann, Organization of metallic nanoparticles using tobacco mosaic virus templates, Nano Lett. 3 (2003) 413–417. https://doi.org/10.1021/nl034004o

[5]   J. Xie, Y. Zheng and J. Y. Ying, Protein – directed synthesis of highly fluorescent gold nanoclusters, J. Am. Chem. Soc. 131 (2009) 888–889. https://doi.org/10.1021/ja806804u

[6]   M. Mertig, L.C. Ciacchi, R. Seidel, W. Pompe and A. De Vita, DNA as a selective metallization template, Nano Lett. 2 (2002) 841–844. https://doi.org/10.1021/nl025612r

[7]   F.A. Armstrong, N.A. Belsey, J.A. Cracknell, G. Goldet, A. Parkin, E. Reisner, K.A. Vincent, A.F. Wait, Dynamic electrochemical investigations of hydrogen oxidation and reproduction by enzymes and implications for future technology, Chem. Soc. Rev. 38 (2009) 36−51. https://doi.org/10.1039/B801144N

[8]   M. Frey, Hydrogenases: hydrogen – activating enzymes, Chem Bio Chem 3 (2002) 153−160. https://doi.org/10.1002/1439-7633(20020301)3:2/3<153::AID-CBIC153>3.0.CO;2-B

[9]   C. Tard, C.J. Pickett, Structural and functional analogues of the active sites of the [Fe]-, [NiFe]-, [FeFe]- Hydrogenases, J. Chem. Rev. 109 (2009) 2245−2274. https://doi.org/10.1021/cr800542q

[10]   X.D. Liu, H.Y. Diao, N. Nishi, Applied chemistry of natural DNA, Chem. Soc. Rev. 37 (2008) 2745−2757. https://doi.org/10.1039/b801433g

[11]   A.J. Patil, J.L. Vickery, T.B. Scott, S. Mann, Aqueous stabilization and self – assembly of graphene sheets into layered bio-nano composites using DNA, Adv. Mater. 21 (2009) 3159−3164. https://doi.org/10.1002/adma.200803633

[12]   H.S. Kim, M. Kim, M.S. Kang, J. Ahn, Y. Sung, and W.C Yoo, Bio-inspired synthesis of melanin-like nanoparticles for highly N-doped carbons utilized as enhanced CO_2 adsorbents and efficient oxygen reduction catalysts, ACS Sustainable Chem. Eng. 6 (2018) 2324-2333. https://doi.org/10.1021/acssuschemeng.7b03680

[13]   Y. Liu, K. Ai, J. Liu, M. Deng, Y. He, L. Lu, Dopamine - melanin colloidal nanospheres: An efficient near-infrared photothermal therapeutic agent for invivo cancer therapy, Adv. Mater. 25 (2013) 1353–1359. https://doi.org/10.1002/adma.201204683

[14]   Y.E. Miao, J. Yan, Y. Ouyang, H. Lu, F. Lai, Y. Wu, T. Liu, A Bio-inspired N-doped porous carbon electrocatalyst with hierarchical superstructure for efficient oxygen reduction reaction, Appl. Surf. Sci. 443 (2018) 266-273. https://doi.org/10.1016/j.apsusc.2018.02.279

[15]   F.A. Armstrong, and J. Hirst, Reversibility and efficiency in electrocatalytic energy conversion and lessons from enzymes, Proc. Natl. Acad. Sci. USA 108 (2011) 14049–14054. https://doi.org/10.1073/pnas.1103697108

[16] Y.Q. Lyu, F. Ciucci, Activating the bio functionality of a perovskite oxide toward oxygen reduction and oxygen evolution reactions, ACS Appl. Mater. 9 (2017) 35829–35836. https://doi.org/10.1021/acsami.7b10216

[17] S. Ghosh and R. N. Basu, Multifunctional nanostructured electrocatalysts for energy conversion and storage: current status and perspectives, Nanoscale 10 (2018) 11241-11280. https://doi.org/10.1039/C8NR01032C

[18] U.A. Paulus, T.J. Schmidt, H.A. Gasteiger and R.J. Behm, Oxygen reduction on a high surface area Pt/ Vulcan carbon catalyst: A thin film rotating ring-disk electrode study, J. Electro anal. Chem. 495 (2001) 134–145. https://doi.org/10.1016/S0022-0728(00)00407-1

[19] B. E. Conway, B. V. Tilak, Interfacial processes involving electrocatalytic evolution and oxidation of H_2 and the role of chemisorbed H, Electrochim. Acta, 47 (2002) 3571–3594. https://doi.org/10.1016/S0013-4686(02)00329-8

[20] M. Tahir, L. Pan, F. Idrees, X. Zhang, L. Wang, J. Zou, Z.L. Wang, Electro catalytic oxygen evolution reaction for energy conversion and storage: A comprehensive review, Nano Energy 37 (2017) 136–157. https://doi.org/10.1016/j.nanoen.2017.05.022

[21] T. Huang, S. Mao, G. Zhou, Z. Zhang , Z. Wen, X. Huang, S. Ci and J. Chen, High performance catalyst support for methanol oxidation with graphene and vanadium carbonitride, Nanoscale 7 (2015) 1301. https://doi.org/10.1039/C4NR05244G

[22] A.S. Arico, S. Srinivasan, and V. Antonucci, DMFCs: From fundamental aspects to technology development. Fuel Cells 1 (2001) 133–161. https://doi.org/10.1002/1615-6854(200107)1:2<133::AID-FUCE133>3.0.CO;2-5

[23] T. Shen, J. Zhang, K. Chen, S. Deng, and D. Wang, Recent Progress of Palladium-based Electrocatalysts for the Formic Acid Oxidation Reaction, Energy Fuels 34 (2020) 9137-9153. https://doi.org/10.1021/acs.energyfuels.0c01820

[24] J.F. Vilaplana, J.V.P. Rondon, C.B. Rogero, J.M. Feliu, E. Herrero, Formic acid oxidation on platinum electrodes: A detailed mechanism supported by experiments and calculations on well-defined surfaces, J. Mater. Chem. A 5 (2017) 21773-21784. https://doi.org/10.1039/C7TA07116G

[25] A. Serov and C. Kwak, Direct hydrazine fuel cells: A review, Appl. Catal. B: Environmental 98 (2010) 1–9. https://doi.org/10.1016/j.apcatb.2010.05.005

[26]   J. Huang, Q. Lu, X. Ma, X. Yang, Bio-inspired FeN$_5$ moieties anchored on three-dimensional graphene aerogel to improve oxygen reduction catalytic performance, J. Mater. Chem. A 6 (2018) 18488-18497. https://doi.org/10.1039/C8TA06455E

[27]   R. Cao, R. Thapa1, H. Kim, X. Xu, M.G. Kim, Q.Li, N. Park, M. Liu, J. Cho, Promotion of oxygen reduction by a bio-inspired tethered iron phthalocyanine carbon nanotube-based catalyst, Nat. Commun. 4 (2013) 2076. https://doi.org/10.1038/ncomms3076

[28]   Z. Tong, D. Yang, X. Zhao, J. Shi, F. Dinga, X. Zou, Z. Jiang, Bio-inspired synthesis of three-dimensional porous g-C$_3$N$_4$@carbon microflowers with enhanced oxygen evolution reactivity, Chem. Eng. J. 337 (2018) 312–321. https://doi.org/10.1016/j.cej.2017.12.064

[29]   C.D. Giovanni, W. Wang, S. Nowak, J. Greneche, H. Lecoq, L. Mouton, M. Giraud, C. Tard, Bioinspired iron sulfide nanoparticles for cheap and long-lived electrocatalytic molecular hydrogen evolution in neutral water, ACS Catal. 4 (2014) 681−687. https://doi.org/10.1021/cs4011698

[30]   Y. Yan, X.R. Shi, M. Miao, T. He, Z.H. Dong, K. Zhan, J.H. Yang, B. Zhao, B.Y. Xia, Bio-inspired design of hierarchical FeP nanostructure arrays for the hydrogen evolution reaction, Nano. Res. 11 (2018), 3537–3547. https://doi.org/10.1007/s12274-017-1919-2

[31]   A. Wang , K. Ju , Q. Zhang, P. Song, J. Wei and J. Feng, Folic acid bio-inspired route for facile synthesis of AuPt nanodendrites as enhanced electrocatalysts for methanol and ethanol oxidation reactions, J. Power Sources 326 (2016) 227–234. https://doi.org/10.1016/j.jpowsour.2016.06.115

[32]   K. Qu, L. Wu, J. Ren, X. Qu, Natural DNA - modified graphene/Pd nanoparticles as highly active catalyst for formic acid electro-oxidation and for the suzuki reaction, ACS Appl. Mater. Interfaces 4 (2012) 5001−5009. https://doi.org/10.1021/am301376m

[33]   D. Grumelli, B. Wurster, S. Stepanow and K. Kern, Bio-inspired nanocatalysts for the oxygen reduction reaction, Nat. Commun. 4 (2013) 2904. https://doi.org/10.1038/ncomms3904

[34]   W. Wei, H.W. Liang, K. Parvez, X.D. Zhuang, X.L. Feng, K. Müllen, Nitrogen-doped carbon nanosheets with size-defined mesopores as highly efficient metal-free catalyst for the oxygen reduction reaction, Angew. Chem. 126 (2014) 1596-1600. https://doi.org/10.1002/ange.201307319

[35] T.N. Ye, L.B. Lv, X.H. Li, M. Xu, J.S. Chen, Strongly veined carbon nano leaves as a highly efficient metal-free electrocatalyst, Angew. Chem. Int. Ed. 53 (2014) 6905-6909. https://doi.org/10.1002/anie.201403363

[36] T.N. Huan, R.T. Jane, A. Benayad, L. Guetaz, P.D. Tran, V. Artero, Bio-inspired noble metal-free nanomaterials approaching platinum performances for H_2 evolution and uptake, Energy. Environ. Sci. 9 (2016) 940-947. https://doi.org/10.1039/C5EE02739J

[37] P.D. Tran, A.L. Goff, J. Heidkamp, B. Jousselme, N. Guillet, S. Palacin, H. Dau, M. Fontecave and V. Artero, Noncovalent modification of carbon nanotubes with pyrene – functionalized nickel complexes: Carbon monooxide tolerant catalysts for hydrogen evolution and uptake, Angew. Chem. Int. Ed. 50 (2011) 1371-1374. https://doi.org/10.1002/anie.201005427

[38] W.J. Shaw, M.L. Helm and D.L. DuBois, A modular energy – based approach to the development of nickel containing molecular electrocatalysts for hydrogen production and oxidation, Biochim. Biophys. Acta Bioenerg. 1827 (2013) 1123-1139. https://doi.org/10.1016/j.bbabio.2013.01.003

[39] A. Lasia, Hydrogen Evolution Reaction, Handbook of Fuel Cells (2010). https://doi.org/10.1002/9780470974001.f204033

[40] Z.S. Cai, Y. Shi, S.S. Bao, Y. Shen, X.H. Xia, L.M. Zheng, Bioinspired Engineering of Cobalt-Phosphate Nanosheets for Robust Hydrogen Evolution Reaction, ACS Catal. 8 (2018) 3895−3902. https://doi.org/10.1021/acscatal.7b04276

[41] S. Cobo, J. Heidkamp, P.A. Jacques, J. Fize, V. Fourmond, L. Guetaz, B. Jousselme, V. Ivanova, H. Dau, S. Palacin, M. Fontecave, V. Artero, A Janus cobalt-based catalytic material for electrosplitting of water. Nat. Mater. 11 (2012) 802−807. https://doi.org/10.1038/nmat3385

[42] N. Jiang, L. Bogoev, M. Popova, S. Gul, J. Yano, Y.Sun, Electrodeposited nickel-sulfide films as competent hydrogen evolution catalysts in neutral water, J. Mater. Chem.A 2 (2014) 19407−19414. https://doi.org/10.1039/C4TA04339A

[43] Y. Sun, C. Liu, D.C. Grauer, J.J. Yano, R.R. Long, P.C. Yang, J. Chang, Electrodeposited cobalt-sulfide catalyst for electrochemical and photo electrochemical hydrogen generation from water, J. Am. Chem. Soc. 135 (2013) 17699−17702. https://doi.org/10.1021/ja4094764

[44] H.J. Cui, Z. Zhou, D.Z. Jia, Heteroatom-doped graphene as electrocatalysts for air cathodes, Mater. Horiz. 4 (2017) 7−19. https://doi.org/10.1039/C6MH00358C

[45]   G.L. Tian, Q. Zhang, B.S. Zhang, Y.G. Jin, J.Q. Huang, D.S. Su, F. Wei, Toward full exposure of "active sites": Nano carbon electrocatalyst with surface enriched nitrogen for superior oxygen reduction and evolution reactivity, Adv. Funct. Mater. 24 (2014) 5956–5961. https://doi.org/10.1002/adfm.201401264

[46]   Z.X. Pei, J.X. Gu, Y.K. Wang, Z.J. Tang, Z.X. Liu, Y. Huang, Y. Huang, J.X. Zhao, Z.F. Chen, C.Y. Zhi, Component matters: paving the roadmap toward enhanced electrocatalytic performance of graphitic C_3N_4-based catalysts via atomic tuning, ACS Nano 11 (2017) 6004–6014. https://doi.org/10.1021/acsnano.7b01908

[47]   S. Chen, J.J. Duan, P.J. Bian, Y.H. Tang, R.K. Zheng, S.Z. Qiao, Three-dimensional smart catalyst electrode for oxygen evolution reaction, Adv. Energy Mater. 5 (2015) 1500936. https://doi.org/10.1002/aenm.201500936

[48]   Z.X. Pei, H.F. Li, Y. Huang, Q. Xue, Y. Huang, M.S. Zhu, Z.F. Wang, C.Y. Zhi, Texturing in situ: N, S-enriched hierarchically porous carbon as a highly active reversible oxygen electrocatalyst, Energy Environ. Sci. 10 (2017) 742–749. https://doi.org/10.1039/C6EE03265F

[49]   T.J. Schmidt, H.A. Gasteiger, R.J. Behm, Methanol electro oxidation on a colloidal PtRu-alloy fuel cell catalyst, Electrochem. Commun. 1 (1999) 1–4. https://doi.org/10.1016/S1388-2481(98)00004-6

[50]   N.M. Markovic and P.N. Ross Surface science studies of model fuel cell electrocatalysts, Jr. Surf. Sci. Rep. 45 (2002) 117–229. https://doi.org/10.1016/S0167-5729(01)00022-X

[51]   J.Y. Kim, T. Kim, J.W. Suk, H. Chou, J.H. Jang, J.H. Lee, I.N. Kholmanov, D. Akinwande, R.S. Ruoff, Enhanced dielectric performance in polymer composite films with carbon nanotube – reduced graphene oxide hybrid filler, Small, 10 (2014) 3405–3411. https://doi.org/10.1002/smll.201400363

[52]   H. Lee, S.E. Habas, G.A. Somorjai, P. Yang, Localized Pd over growth on cubic Pt nano crystals for enhanced electro catalytic oxidation of formic acid, J. Am. Chem. Soc. 130 (2008) 5406−5407. https://doi.org/10.1021/ja800656y

[53]   S. Zhang, Y. Shao, G. Yin, Y. Angew, Electrostatic self – assembly of a Pt around Au nanocomposite with high activity towards formic acid oxidation, Angew. Chem., Int. Ed. 49 (2010) 2211−2214. https://doi.org/10.1002/anie.200906987

[54]   I.V. Yang, H.H. Throp, Kinetics of metal-mediated one electron oxidation of guanine in polymeric DNA and in oligonucleotides containing trinucleotide repeat sequences, Inorg. Chem. 2000, 39, 4969−4976. https://doi.org/10.1021/ic000607g

Bioinspired Nanomaterials for Energy and Environmental Applications　　　Materials Research Forum LLC
Materials Research Foundations **121** (2022) 117-140　　　https://doi.org/10.21741/9781644901830-4

[55]　M.E Napier,　D.O. Hull, H.H. Thorp, Electrocatalytic oxidation of DNA –
　　　wrapped carbon nano tubes, J. Am. Chem. Soc. 2005, 127, 11952−11953.
　　　https://doi.org/10.1021/ja054162c

Bioinspired Nanomaterials for Energy and Environmental Applications   Materials Research Forum LLC
Materials Research Foundations **121** (2022) 141-174   https://doi.org/10.21741/9781644901830-5

Chapter 5

Bioinspired Nanomaterials for Supercapacitor Applications

Adhigan Murali[1,*], R. Suresh Babu[2], M. Sakar[3], Sahariya Priya[4], R. Vinodh[5], K. P. Bhuvana[1], Senthil A. Gurusamy Thangavelu[6], Aashish S. Roy[7], M. Abdul Kader[1]

[1]School for Advanced Research in Polymers (SARP)-ARSTPS, Central Institute of Plastics Engineering & Technology (CIPET), Chennai-600032, India

[2]Laboratory of Experimental and Applied Physics, Centro Federal de Educação Tecnológica, Celso Suckow da Fonseca, Av. Maracanã 229, Rio de Janeiro, 20271-110, Brazil

[3]Centre for Nano and Material Sciences, Jain University, Bangalore 562112, Karnataka, India.

[4]Department of Plant Biology & Biotechnology, Loyola College, Chennai-600034, India

[5]School of Electrical and Computer Engineering, Pusan National University, Busan- 46241, Republic of Korea

[6]SRM Research Institute and Department of Chemistry, SRM Institute of Science and Technology, Kattankulathur, Chennai 603203, India

[7]Department of Chemistry, S.S.Tegnoor Degree College, Kalaburagi, Karnataka, India

* precymurali@gmail.com

Abstract

Energy storage devices have acquired great research attention in the fabrication of ultra-high efficient supercapacitors. In order to enhance the electrochemical performance of the supercapacitors, different electrodes have been fabricated using various nanomaterials with precisely controlled morphologies and interfaces. Nevertheless, the low-dimensional nanomaterials still suffer from the factors such as severe re-stacking, non-homogeneous aggregation, and low contacts during the processing and assembly. These bottle-neck problems essentially lead to the hindrance of transport of electrons and/or ions in the energy devices. In this direction, recently, the bioinspired nanomaterials are emerging as the potential candidates to overcome the said disadvantages of the chemically derived low dimensional nanomaterials. The well-aligned or highly oriented bioinspired nanostructures found to effectively promote the transport of electrons, facilitate the ion diffusions through the hierarchical pores and provide the large specific surface area for their interfacial interactions with the surroundings. Moreover, the nanoscale materials can be easily tuned or engineered for their physicochemical properties, thereby they can be potentially used in

Bioinspired Nanomaterials for Energy and Environmental Applications   Materials Research Forum LLC
Materials Research Foundations **121** (2022) 141-174   https://doi.org/10.21741/9781644901830-5

many device applications. In this context, this chapter is intended to highlight the recent progress in bioinspired nanomaterials towards developing the electrode materials for supercapacitors with the emphasize on the fundamental understandings between their structural properties and electrochemical performances. Finally, it concludes with an outlook on the next generation nanostructured electrodes to design the ultra high-efficient supercapacitors.

Keywords

Bioinspired Nanomaterials, Energy Storage, Supercapacitors, Biomass, Protein Nanotubes, Graphene

Contents

Bioinspired Nanomaterials for Supercapacitor Applications **141**

1.   Introduction...**143**

2.   Nanostructured materials for electrostatic double-layer capacitors (EDLCs) **145**

    2.1   Carbon nanotubes (CNTs) 145

    2.2   Single-walled carbon nanotubes (SWCNTs)...................................145

    2.3   Bioinspired CNTs and graphene in Electrostatic double-
           layer capacitors (EDLCs) 146

    2.4   Bioinspired peptide based nanotubes electrodes for EDLC............150

    2.5   Biomass-derived carbon 151

3.   Nanostructured materials for pseudocapacitors **153**

    3.1   Metal hexacyanoferrates..153

    3.2   Metal oxides...155

    3.3   Ruthenium oxide (RuO$_2$) 156

    3.4   Manganese oxide (MnO$_2$)...156

    3.5   Cobalt oxide (CaO)..157

    3.6   Other MOs 157

    3.7   Transition Metal Sulfides (TMS) 158

    3.8   Bioinspired conducting polymers (CPs) nanostructures 159

Concluding remarks and future prospects...**160**

Acknowledgements...**161**

References **161**

1.   Introduction

The depletion of non-renewable fossil fuel resources such as oil and coal has caused a severe crisis worldwide in energy productions. [1-3]. Therefore, there is an increasing attraction towards find the alternative and sustainable materials to meet the energy demands. Accordingly, there is an increasing demand in order to develop the hybrid electric vehicles and several portable electronic devices with high energy storage performances. Particularly, the battery devices are facing various complications during their charge and discharge operations. On the other hand, the supercapacitor, also known as ultracapacitor or electrochemical capacitor, can be an excellent device for energy storage. Recently, there are a large number of research works carried out by the scientist/researchers in order to improve the energy storage efficiency of the supercapacitors towards enhancing their performances in energy applications. In this direction, the typical physicochemical properties of different nanomaterials have been effectively exploited to develop the energy storage devices such as supercapacitor and batteries [4,5].

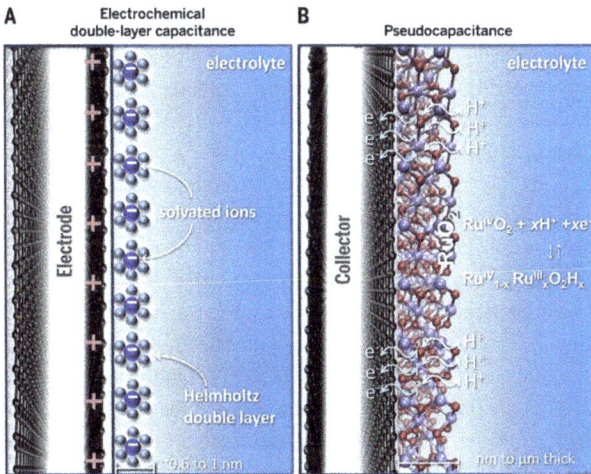

Figure 1. Schematic representation of the supercapacitors (A) adsorption of ions at EDLC surface, (B) charge transfer at the electrode surface (pseudocapacitor). Reproduced with permission from Ref no [7].

Bioinspired Nanomaterials for Energy and Environmental Applications   Materials Research Forum LLC
Materials Research Foundations **121** (2022) 141-174   https://doi.org/10.21741/9781644901830-5

Basically, supercapacitors can be classified into two main categories based on their energy storage mechanisms [6,7] as shown in Fig. 1. The first mechanism involves a pure electrostatic interaction between the electrolytic ions and the charged electrode surface, which is known as the electrochemical double layer capacitance (EDLC) and on the other hand, the energy storage mechanism can happen through Faradaic redox reaction called as pseudocapacitors. In EDLC, various carbonaceous materials such as graphene, activated carbon and carbon nanotubes (CNTs) are used as electrode materials. Similarly, in the pseudocapacitive systems, the materials such as transition metal hexacyanoferrates, metal oxides and conducting polymers are used to fabricate their electrodes. It should be noted that the types of electrode materials play a very important role in determining the electrochemical operations in supercapacitor devices [8]. In order to evaluate the performances of a supercapacitor, the parameters such as cycling stability, specific capacitance and electrolyte should be estimated. Also, there are important parameters such as specific capacitance, pore size morphology, ionic and electronic conductivity [9] also involve in enhancing the performances of the supercapacitor. Generally, pseudocapacitive materials exhibit the high specific capacitance and a variety of carbonaceous materials have exhibited a superior cyclic stability up to 100,000 cycles. However, supercapacitors still face many problems such as low electrical conductivity, weak structural stability and charge carrier mobility [10], which can be effectively achieved using nanoscale materials to fabricate their electrodes. The formation of robust active sites and defects in the grain boundaries in nanomaterials lead to the emergence of multiple redox activity, superior ionic conductivity and short diffusion pathway. Accordingly, there are different types of spinal nanostructures, metal oxide, core-shell, redox polymers or conductive polymers such as polyaniline, polyimide, acrylates and acrylic polymers can be effectively used and they can be easily synthesized in various ways including hydrothermal, radical polymerization and high temperature sintering process [11, 12]. Hence, the nanoscale hybridization of EDLC and pseudocapacitive electrode materials of different dimensionalities such as 1D, 2D and 3D materials can be used as they possess the increased high surface area, where they can enhance the electrochemical properties of supercapacitors. In this direction, nanostructures with desired properties and functions can be synthesized via biologically inspired routes using various biological structures and biomolecules. Notably, these bioinspired materials will also have the unique physico chemical properties of their respective biological entities as well [13,14]. It is found viable that the supercapacitors can be industrialized in pilot scale with careful choice of electrode materials synthesized from biological materials, which are basically inexpensive and involve simple preparation procedures. Hence, there are important investigations are being progressed in order to explore the new design of electrode materials [15].

Bioinspired Nanomaterials for Energy and Environmental Applications   Materials Research Forum LLC
Materials Research Foundations **121** (2022) 141-174   https://doi.org/10.21741/9781644901830-5

2. Nanostructured materials for electrostatic double-layer capacitors (EDLCs)

2.1   Carbon nanotubes (CNTs)

Carbon nanotubes were discovered by Sumio Iijima in 1991 using arc-evaporation method [16]. However, CNTs were discovered long before in the form of allotropes of carbon with diameter range between 1 and 100 nm. It is basically an artificial allotrope of carbon containing the single or multiple layers of graphene rolled up to form a cylindrical nanostructure, also known as tubular structure of fullerenes, which is fully composed of sp^2 hybridised carbon. Generally, CNTs can be classified into three types namely, single-walled carbon nanotubes (SWCNTs), double-walled carbon nanotubes (DWCNTs) and multi-walled carbon nanotubes (MWCNTs). SWCNTs are made of single layer of graphene sheet rolled up with a diameter of 102 nm. Similarly, the multi-walled carbon nanotubes consist of multiple layers of graphene sheet rolled upon itself with diameter range between 2 to 50 nm. Since last decades, CNTs are synthesized in different techniques such as arc discharge, laser ablation, high pressure carbon monoxide (HiPCO) and chemical vapor deposition (CVD) [17-21]. Usually, CNTs contain a mixture of impurities such as amorphous carbon, carbonaceous materials, metallic catalyst and multi-wall graphite. There are different purification methods established to remove those impurities through acid oxidation and high temperature heat treatment in different atmospheric conditions [22]. CNTs have excellent distinctive chemical, mechanical, electrical, thermal and optical properties. Moreover, it is 100 times stronger than the steel because of their perfect intertube arrangement with covalently connected -C-C bonding [23, 24].

2.2   Single-walled carbon nanotubes (SWCNTs)

Single-walled carbon nanotubes are 1-dimentional (1-D) quantum structure or pseudo-quantum wire, which consists of sp^2 hybridized carbon [25]. However, tube diameter of the carbon are very important to tune their unique physical and chemical properties, which are highly related to the high electrical conductivity, surface area, thermal stability and metallic to semi metallic current carrying capacity [26]. When compared to MWCNTs, the SWCNTs acquire extraordinary mechanical properties such as high Young's modulus, which estimated to be around 1-5 TPa, high thermal conductivity, high tensile strength (50-200 GPa) and high aspect ratio [27,28]. Also, they consist of topological defects due to the presence of pentagons or heptagons at the carbon framework and curvature sites, rehybridization (sp^2-sp^3), incomplete bonding defect between vacancies and dislocations of their surfaces. SWCNTs are difficult to disperse, which lead poor solubility in all organic solvents due to the presence of strong intermolecular cohesive force (0.5 eV nm^{-1}) between tube-to-tube interactions [29].

Different types of carbonacious materials such as amorphous carbon and graphitic metal nanoparticles are found to be as impurities in the pristine carbon nanotubes. In order to purify the CNTs, following methods have been involved to remove the impurities from CNTs: gas-or vapour phase oxidation, centrifugation, thermogravimetric analysis (TGA) and wet chemical oxidation. Similarly, many strategies were developed in order to debundling of SWCNTs *via* covalent and non-covalent modifications. Covalent functionalization induce some damages on surface of SWCNTs, whereas, the non-covalent modifications enhance the dispersion with less damages on SWCNTs [30].

2.3   Bioinspired CNTs and graphene in Electrostatic double-layer capacitors (EDLCs)

In 1997, Niu et al., [31] suggested that CNTs can be used in supercapacitors. Particularly, the functionalized CNTs can have some of the functional groups on their surfaces, as because these functionalized CNTs possess specific area of around 430 m^2/g with a gravimetric capacitance of around 102 F/g and an energy density of around 0.5 Wh/kg. As discussed before, residues of CNTs can be removed using acid oxidation or calcinations. In this way, around 90% of catalyst residues were removed before the electrochemical process and the remaining 10 % catalyst was found to involve in the Faradiac and non-Faradiac reactions in supercapacitor. The CNTs are used in supercapacitors due to their functional groups as well as catalytic behavior. The structural properties of CNTs such as diameter, length and pore sixe also play a vital role on the EDLCs [32,33]. Romano et al., also reported the graphene based carbon nanotubes for EDLC application with ultra areal performance [34]. The nanostructured allotropes of carbon materials such as activated carbons, carbide derived carbon, graphene and carbon nanotubes are greatly demonstrated in order to use in EDLC systems. Mostly, the commercial EDLC devices consist of the activated carbons, because of their high surface area (3000 m^2g^{-1}) and low cost. Usually, the activated carbon films show easy accessibility with electrolyte ions, which hinder the maximum power density. Among the various CNTs, the single walled/double walled carbon nanotubes were active materials due to their high specific capacity, where it is predicted that the theoretical specific capacity of single walled carbon nanotubes is around 1315 m^2g^{-1} and for the double walled it is around 800 m^2g^{-1} [35,36]. Rakesh et al., reported the fabrication of electrochemical double layer capacitor electrodes using carbon nanotubes by air assisted chemical vapor deposition technique [37].

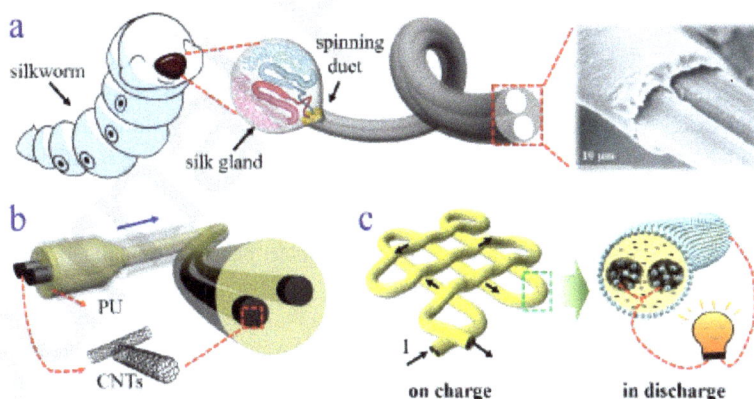

Figure 2. Schematic representation of bioinspired CNTs for supercapacitor, Reproduced permission from Ref no [38].

Guo et al., reported the bioinspired multicomponent carbon nanotubes, which are synthesized from microfluidics for supercapacitor applications. Usually, the fiber based supercapacitors are being considered to be one of the potential power storage devices because of their light weight, high flexibility and wear ability. In this study, they have developed the silkworms based nanofiber structure by co-flow microfluide spinning method with combination of polyurethane and CNTs. These materials showed excellent flexibility, stability and better cycle life performance, which can be used for better electrochemical energy storage applications [38]. Generally, silkworms such as Bombyx mori are silk proteins, which can produce directly into silk fibers that have more core shell structures on their surface. Still there are lack of bioinspired nano and biomaterial for the usage of supercapacitor or battery applications. Hence, Guo and co-authors focused on the Bombyx mori silk fibers with combination of carbon nanotubes for supercapacitor applications as shown in Fig 2. The outcome this work involved the development of silk fiber based CNT/PUs with excellent power density with reasonable cyclic stability, which could be used for EDLC. The typical capacitance of the developed nanofibrous structures was estimated to be 231.89 C/m^2 and $1.34*10^5$ C/m^3 for 440 μm thickness of fibres. Moreover, their specific areal- and volumetric- capacitances were around 23.19 mF/cm^2

and 134 mF/cm^3, respectively. Hence, these biologically inspired nanomaterials were suggested to be good candidates for various supercapacitor applications. Guoping and co-authors developed the bioinspired leaves-on-branchlet structure consisting of carbon nanotubes serving as branchlets and graphene petals as leaves for electrodes. The developed bioinspired carbon materials exhibited high areal capacitance of 2.35 Fcm^{-2}, high rate capability and outstanding cyclic stability (10000 cycles) [39]. Generally, CNTs array is most promising materials with better rate capability, cyclic stability and energy density. However, the common issues associated with CNTs array electrodes include the poor bonding of nanotubes to substrate, low tube-to-tube charge transfer efficiency and easy destruction of the tube orientation. In this reported structure, as the graphene petals possess high surface area, sharp edges and high electrical conductivity, these bioinspired graphene and CNTs were found to be promising electrochemical candidate and ideal structures for supercapacitor applications [40-44]. The nanostructured porous materials inspired from different biological structures can be derived from waste biomasses, polymers or inorganic chemical, which could be renewable and environmentally benign for supercapacitors. In this direction, the two step microwave plasma chemical vapor deposition (MPCVD) method was adopted to synthesize the highly porous nanostructured electrode inspired by nature of tree branchlets. Graphene petals decorated with highly oriented carbon nanotubes array were developed for EDLCs. The schematic representation of overall synthesis strategy for Co_9S_8@CNT/CNF composite was shown in Fig.3. Also, TEM images are clearly showed that the CNT diameter of about 200 nm are twisted by the CNF and moreover, the cyclic voltammetry with different scan rate (2 to 60 mV S^{-1}) clearly showed their better electrochemical behavior. There are two pair of redox peaks, which indicated that the redox reaction take place in Co_9S_8 electrode and electrolytes, when sweap rate increased, there is no current peak was observed, which suggesting that the Co_9S_8@CNT/CNF electrode was favourable for fast redox reactions [39b]. The hierarchical carbon micro conduits with hallow channels were designed in order to increase the surface area of the electrolyte, which basically access the ion diffusion during electrochemical performance in supercapacitors. This fabricated electrode showed high areal capacitance, good rate capability and excellent cyclic stability [39]. Lilli et. al., reported the bio-derived hierarchical carbon-graphene based TiO_2 for supercapacitor applications [45]. They have synthesized the carbon-graphene based TiO_2 using loofah scaffold with enhanced electron transfer efficiency and improved conductivity. Generally, carbon materials such as CNTs, activated carbon and graphene possess high surface area, which in turn decrease the volumetric capacitance due to thier high porosity.

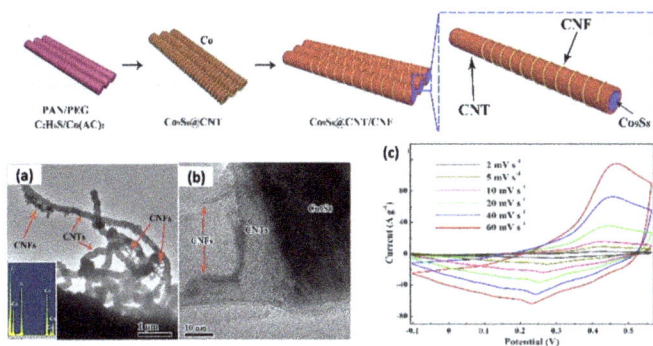

Figure 3. Schematic representation of Co_9S_8@CNT/CNF; (a, b) TEM images of Co_9S_8@CNT/CNF, and (c)Cyclic voltammetry curve for different scan rate. Reproduced permission from Ref no [39b].

Moreover, the carbon based materials were mostly derived from biomaterials such as fruit peels, bamboo and seaweed by carbonization, which provided the excellent specific surface area and superior electrochemical stability for EDLC [46-48]. Jacob et al. reported the EDLC developed using ionic liquid 1-butyl-1-methylpyrrolidinium tris (pentafluoroethyl) trifluorophosphate as an electrolyte with combination of graphene as electrodes, which showed excellent high energy density of 25 Wh/kg along with high specific capacitance 200 F/g at >5 operating voltage window. In the past decade, carbon based materials integrated with 2D materials, MOF, redox active polymers have also been tested for EDLC applications. Among them, graphene, CNTs, activated carbon and hybrid structures are preferred electrode materials for EDLC, because of their high conductivity, low reactivity and high specific surface area [49]. Singh et al., reported the nanoflower like architecture decorated by 3D graphene based MoS_2 on graphite electrode in order to design a solid state supercapacitor device. The specific capacitance was obtained to be around 58.0 F and the energy density of 24.59 Wh/Kg with power density 8.8 W/Kg [50]. It has been also reported that the capacitance is found increasing because of the increase in Fradaic peaks in EDLC. Ko et. al reported the porous graphitic carbon with $Ni_2P_2O_7$ heterostructure, which showed high specific capacitance around 1893 F/g [51]. Liao et al., reported an ultra high capacitance of 3480 F/g using Co_3O_4 nanoparticles onto graphene sheet. However, the different solid state energy storage devices showed some disadvantages such as lack of stability and poor power density. More importantly, the crystallinity and morphology (porous) play a crucial role in specific capacitance [52]. In this context, Duc et al., reported that the bioinspired sponge carbon immobilized graphene, which demonstrated for an enhanced flexible transparent supercapacitor performances [53]. In their study, they used the CVD method to produce sponge like carbon materials. The leaf-skeleton inspired electrodes were used to provide a 3D spongy core. The sheet contained 88% optical transmittance, an electrical sheet resistance of 1.8 Ω/sq and areal capacitance was around 7.06 mFcm^{-2}.

2.4   Bioinspired peptide based nanotubes electrodes for EDLC

Bioinspired nanomaterials exhibit unique physical and chemical properties which cannot be simply achieved by biological entities. Supercapacitors basically possess lower energy density than batteries, hence, the there are considerable attention being focused in order to develop a high surface area based electrolyte and electrode. In this direction, nanomaterials such as nanotubes, nanoparticles and nanowires have become more important materials for electrode applications, which can be used for the electrochemical energy storage devices [54]. Mostly, the carbon based materials have been used in this area because of their high surface area and high energy density. [55]. Accordingly, the modified carbon electrode with diphenylamine peptide nanotubes exhibited enhanced efficiencies. Moreover, these bioinspired peptide nanotubes offer a new route in the developing the nanostructured materials. Natural and non-natural amino acid based structures are found to be potential candidates in self assembling into the hierarchical nanostructures such as nanotubes, nanofibers and nanosphere. These bioinspired materials are expected to have great potentials in next generation wearable electronics. The utilization of peptide based materials can be effectively used for batteries and supercapacitors. Beker et al., [56] reported the electrochemical behavior of charge and discharge in EDLC using peptide nanotubes, which significantly increase their capacitance and surface area. The size of the prepared materials was found to be below 300 nm with wall thickness of around 50 nm and average diameter of around 200 nm along with better porous structures. These nanotube arrays showed a high density of 4×10^8 cm^{-2}. It was found that the carbon material coated with peptide nanotubes with different thickness of peptide nanotubes showed better electrochemical properties [54]. The same group also reported the self assembled bioinspired peptide nanotubes by PVD method for the improved electrochemical energy storage applications. This peptide nanotubes (PNTs) showed improved wettability in electrolytes, which are critical factors for strong variation in the EDLCs. Moreover, the prepared PNTs showed 800 mF/cm^2 in sulfuric acid and drastically increased functional area of carbon electrode. These bioinspired materials were chemically synthesized using biomolecules such as amino acid, peptide, proteins and DNA [56]. More importantly, the first man-made peptide based nanotubular structure was developed by self assembly of flat and cyclic peptide, which contained the D- and L- amino acid residues [57]. Wang et al., reported the bioinspired high performance electrochemical supercapacitor, which contained the conducting polymer modified monolithic carbon. The conducting polymers such as polyaniline, polythiophene and polypyrrole were deposited on monolithic porous carbon and it showed high surface area of 1125 m^2g^{-1}. The specific capacitance was estimated to be 1488 F g^{-1} in sulfuric acid electrolyte. As a result, the overall power density was reached to 49.5 Wh kg^{-1} hence, such a bioinspired materials can be reliably used in electrochemical energy storage devices [54].

2.5   Biomass-derived carbon

Biomass is one of the renewable resources, which is easily available resource on the earth. It has been used for different applications, which include water treatment, dye sensitized solar cells, hydrogen storage and energy storage, owing to their low cost, unique structure, sustainability, renewable and easily availability. The well oriented channels and abundance of the carbon materials are being used for the biological process, which make these biomass derived electrode materials viable for electrochemical storage devices. The synthesis of biomass derived carbon materials has drawn great attention in energy conversion and storages (see Figure 4).

Figure 4. Different types of biomass converted onto energy storage. Reproduced permission from Ref no [64].

These biomass derived carbon nanomaterials offer a new approach for value added utilization of biomass for different energy storage device applications. There are different carbonizations techniques. For example, the hydrothermal and pyrolysis can be used to convert the biomass into activated carbon by tuning their surface properties, morphologies, porosity by controlling various parameters such as temperature, time and reagents [58-63].

Lu et al., reported the cellulose derived microcrystalline, porous and nitrogen doped carbon nanomaterials for electrochemical energy storage applications (see Fig 5a,b) [65,66]. Owing to thier high precentage of nitrogen and porous structure, they were completly exploited for thier charge storage through pseudocapacitive and EDLCs mechanisms. These biomass-dervied carbon nanostructures demonstrated the high specific capacitance of 426 F g^{-1} at 0.25 A g^{-1} current density. Liu et al., prepared the 3D carbon nanosheets with high percentages of N and O doped heteroatoms and were used for EDLCs [67]. In their study, they used the hemi cellulose, which was hydrolyzed by KOH, while hydrogen peroxide was set to continuously oxidize the lignin. In order to incorporate the nitrogen, the urea was used as precursors to introduce N atoms into the carbon sheets. This biomass-carbon materials showed the capacitance of 508 F g^{-1} at 1 A g^{-1} current density and long cyclic stability around 12000 cycles [66]. Interestingly, the dual biomass such as rice straw and glucose derived carbon materials contained the nano and micro carbon sphere, which further attached onto the porous carbon containing substrate and demonstrated for the high performance supercapacitor electrodes [67]. The synthesized materials exhibited the hierarchical porous structure with high surface area of 1122 m^{2}g^{-1} and high percentage of oxygen (14.2 %). The assembled electrodes in the coin cell with 6.0 M KOH as electrolyte showed the increased symmetrical rectangular CV profile (Fig. 5a) with increasing scan rate up to 100 mVs^{-1}. This observed result indicated that this material system can be effectively can be used for EDLC. Similarly, from the charge–discharge studies (Fig 5b), the single electrode of the coin cells found to enhance the specific capacitance of 238 F g^{-1} at 1 Ag^{-1} current capacity with high retention of around 215 F g^{-1} (see Fig 5c). The prepared material also showed the excellent cycling stability in long term.

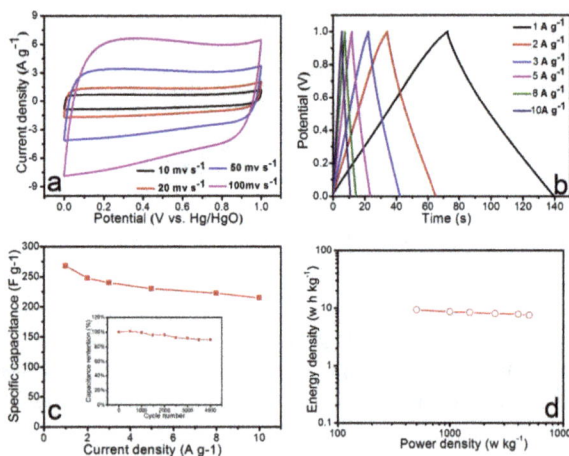

Figure 5. Electrochemical behavior of nano-micro sphere @ rice straw derived porous carbon composites materials (NMCSs@RSPC) in a symmetric type of coin cell: (a) CV profile at various sweep rate, (b) GCD profile at various current densities, (c) specific capacitance for single electrode, (d) regone plot cell. Reproduced permission from Ref no [67].

Bioinspired Nanomaterials for Energy and Environmental Applications   Materials Research Forum LLC
Materials Research Foundations **121** (2022) 141-174   https://doi.org/10.21741/9781644901830-5

In addition to this, various biomass sources were eventually considered as a promising source to prepare the electrodes for supercapacitor as listed in Table 1. As a result, biomass derived carbon materials with better electrochemical behavior were developed via different methods.

Table 1 Comparison of various biomass-derived porous carbon electrodes materials for supercapacitors.

Biomass	Electrolyte system	Cut off window (V)	Specific capacitance ($F\ g^{-1}$)	Current density ($A\ g^{-1}$)	Ref.
Coconut-shell	6.0 M KOH	-0.8-0.2 (Hg/HgO)	192	1.0	[68]
Willow catkin	1.0 M H_2SO_4	-1.0-0 (Ag/AgCl)	298	0.5	[69]
Bamboo	6.0 M KOH	0-1 (Ag/AgCl)	301	0.1	[70]
Banana peel	6.0 M KOH	-1.0-0 (SCE)	206	1.0	[71]
Lignin	6.0 M KOH	-1.0-0 (Ag/AgCl)	312	1.0	[72]
Rose	6.0 M KOH	-1.2-0 (SCE)	208	0.5	[73]
Wallnut shell	6.0 M KOH	-1.0-0 (SCE)	263	0.5	[74]
Cigarette filter	6.0 M KOH	-1.0-0 (Hg/HgO)	263	1.0	[75]

Tian et al., demonstrated the bioinspired beehive-like hierarchical nonporous carbon (BHNC) derived from the industrial waste of bamboo based byproducts. The obtained material showed excellent electronic conductivity (4.5 S cm^{-1}) and high specific surface area (1472 $m^2\ g^{-1}$) followed by chemical activation and calcinations processes. The developed BHNC electrode showed the enhanced high specific capacitance of 301 F g^{-1} at 0.1 A g^{-1} current density. Further, the energy density was estimated to be 43.3 Wh kg^{-1} at maximum power density of 42000 W kg^{-1}, which was obtained in the ionic liquid electrolyte based system [70,71].

3. Nanostructured materials for pseudocapacitors

3.1 Metal hexacyanoferrates

Prussian blue (PB) and their analogues such as transition metal (II) hexacyanoferrate (III) (MHCF) is a kind of polynuclear mixed valence inorganic coordination complex and their common formula is $A_xM_yFe(CN)_{6.z}H_2O$, where A is the counter cation, M reveals the transition metals (M=Fe, Co, Ni), x,y and z are stoichiometry coefficients. The MHCFs typically form as a 3D structured cubical network with the repeating units of -N≡C-Fe-C≡N-M≡N-C- as shown in Fig. 6.

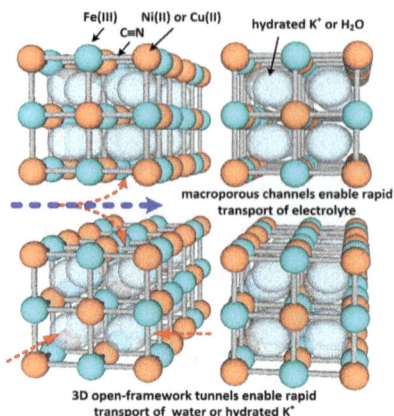

Figure 6. Schematic representation of 3D framework of hexacyanoferrate with hydrated water molecule coated on stainless steel. Reproduced permission from ref. no [76].

MHCFs allow the counter ions insertion/desertion into channels of 3D network, which in turn permit the reversible reduction and oxidation reactions. Owing to their structural characteristics, they show the interesting reversible redox reaction, outstanding stability and environmental safety. Moreover, MHCFs show wide range of applications such as electro-analysis, electrochemical sensor, batteries, ion exchange and supercapacitors [77-80].

Different types of nanoparticles such as insoluble iron, cobalt and nickel hexacyanoferrates nanoparticles were synthesized by cost effective solgel/co-precipitation method suitably for the electronic applications including supercapacitor. Zhao et al., [78] developed an asymmetric supercapacitor based on CoHCF nanoparticles, which also acted as an electrode and showed the highest specific capacitance of 250 Fg^{-1} with extraordinary rate capability and better long-term cyclic stability along with good capacitance continuance of 93.5% even after 5000 cycles [79]. The hybrid MHCFs namely Ni-CoHCF based nanostructures and the hierarchical hybrid CO-MnHCF electrodes were showed unique electrochemical behaviour as compared to the single-MnHCFs electrodes. Moreover, the main disadvantage of MHCF was realized to be the decreasing dissolution during the electrochemical potential cycling process [80-82].

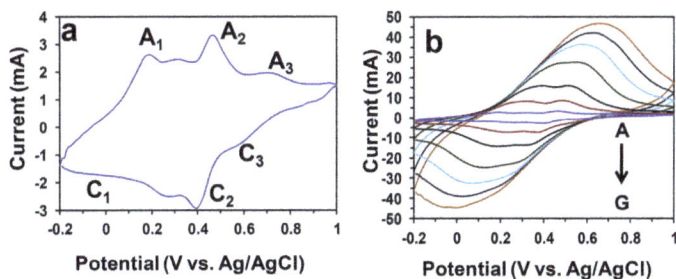

Figure 7. CV of PANI/MnHCF electrode in equal volume of H_2SO_4 and Na_2SO_4 mixed electrolyte at the sweep rate of 2 mVs^{-1} and (b) CV of PANI/MnHCF electrode at various sweep rate from 2 to 50 mVs^{-1} (A-G). Reproduced with permission from [82].

In order to resolve this problem, there were several ways developed to enhance their stability and specific capacitance. More particularly, the conducting polymers such as polyaniline with various MHCFs were tested to improve their electrochemical behavior. A highly flexible and transparent carbon fibre electrode was coated with polyaniline/manganese hexacyanoferrate (PANI/MnHCF), where it showed the excellent redox reaction and high specific capacitance with excellent stability as shown in Fig 7.

3.2   Metal oxides

The electrodes material such as carbon-based materials, transition metal oxides (TMOs) and conducting polymers (CPs) are often being used for supercapacitor applications. This include the carbon-based materials such as carbon nanotubes (CNTs), single walled carbon nanotubes (SWCNTs), multiwalled carbon nanotubes (MWCNTs), activated carbon (AC), and graphene, which are commonly used as electrode material for EDLC and pseudo capacitors. On the other hand, transition metal oxides (TMOs) tend to enhance the redox reactions in the system. Hence, in recent decades, the TMOs are effectively used as electrode material in pseudocapacitors. Moreover, the TMOs have gained importance as an efficient electrode material for energy storage applications due to their abundance on earth, environmentally friendly in nature and easy accessibility. These TMOs based materials show versatile characteristics such as surface morphology (different structures), high surface area and large specific capacitance. In addition, TMOs also play a crucial role in the electrochemical supercapacitors as an electrode material to deliver a prominent specific capacitance, which associated with their defects under a certain condition. There are certain important factors influencing the electrochemical performance of the TMOs based electrodes are, (i) different structural morphologies, (ii) combination of different metal oxides contributing the synergistic effect in the electrode material, and (iii) tunable redox reaction conditions [84]. Accordingly, it has been found that various TMOs and metal hydroxide provide better faradic reaction and can produce excellent specific capacity and good energy densities. Moreover, it has been widely explored as supercapacitive materials owing to their high energy and power output through nanostructures engineered materials [85]. Metal oxides (MOs) with various chemical compositions such as simple, double or mixed oxides and hybrid composites have also been studied in high-performance electrode properties used for pseudocapacitive systems. Accordingly, some of the widely explored MOs include RuO_2, MnO_2, SnO_2, V_2O_5, Fe_2O_3, Fe_3O_4, Bi_2O_3, WO_3 and In_2O_3 [86-98].

3.3   Ruthenium oxide (RuO_2)

RuO_2 is considered as one of the excellent electrode materials in supercapacitors due to its good theoretical specific capacitance, reversible redox reaction and good electrical conductivity. Further, RuO_2 shows an enhanced corrosion resistance property both in acid and basic conditions in electrochemical reactions. Also, it is a promising material in the applications of various aqueous electrolytes. There are two phases of ruthenium oxide with one having the amorphous hydrous $RuO_2.XH_2O$ and the other one is the crystal phase $RuO_2.XH_2O$. These two kinds of structures have a high specific capacitance and superior active sites. Buzzanca et al. demonstrated first hydrous RuO_2 and used as pseudocapacitive electrode material in an acidic electrolyte [98]. Yet, the main challenges of the RuO_2 electrode, having severe agglomeration during cycling capacitive. Cao et al. synthesized a unique nanostructure material by immobilizing the RuO_2 nanoparticles (diameter of 1.9 nm) on the graphene and carbon nanotubes. These nanocomposites materials were efficiently reduce the agglomeration phenomenon, which can be effectively used in capacitor applications [99].

3.4   Manganese oxide (MnO_2)

Manganese oxide nanoparticles are another important material, which can be used as a potential cathode material for asymmetric supercapacitors. Among the different manganese oxide phases, the MnO_2 and Mn_3O_4 phases have gained more attention for broad application due to their wide working potentials, low cost, easily availability and abundant in nature. Several efforts have been put forth on crossing the gap between theoretical capacitance and experimental capacitance. Zhang et. al. fabricated a core-shell structure based electrode consisting of a B-MnO_2 core and highly aligned "birnessite-type" MnO_2 shell [101]. The fabricated asymmetric device provided a high energy density around 40.4 W H Kg^{-1} with maximum power density of 17.6 KW Kg^{-1}. Recently, many groups have reported that the incorporation of high content of cations such as Na^+ and K^+ into MnO_2 structure can improve their electrochemical performance, based on the additional redox reaction [103,104]. Xia et al. synthesized the carbon-coated Fe_3O_4 nanorod arrays, which can act as anode and asymmetric device with $Na_{0.5}MnO_2$ nanowall arrays as a cathode. These materials exhibited an outstanding energy density of 81 Wh Kg^{-1} in as extensive potential window of -1.3 to 0 V vs Ag/AgCl [103].

3.5   Cobalt oxide (CaO)

Among the variety of metal oxides for supercapacitor applications, the Co_3O_4 based cathode materials were found to be controlled by the diffusion in the electrochemical process in an aqueous electrolyte and considered as battery-type of electrode materials, where these materials were also used in lithium ion and solid state batteries. The Co_3O_4 cathodes were found to be energetic due to their theoretical ultrahigh specific capacitance and excellent redox reversibility. However, as the rate capability and cycling stability are associated by their low electrical conductivity, there was considerable attention focused on the designing of nanostructured Co_3O_4 to enhance the electrochemical performances [105]. Yu et al. reported the Co_3O_4/N-doped carbon hollow spheres with a hierarchical nanostructure and used as an electrode material for a high performance asymmetric supercapacitor device. Upon the assembly of the material as a cathode to fabricate an asymmetric device, the device showed a high energy density of 34.5 Wh Kg^{-1} with a maximum power density of 753 W Kg^{-1} [104].

3.6   Other MOs

Towards increasing the performance of the energy storage systems, several other metal oxides such as MoO_3, WO_3, PbO_2, SnO_2 and CuO have also been explored to develop the electrode materials for supercapacitor applications. Among them, nickel oxide (NiO) is yet another potential anode material for supercapacitors [106-110]. Wei et al. fabricated the honeycomb-like mesoporous NiO via a hydrothermal reaction and used as a high-performance cathode for asymmetric supercapacitor [110]. The assembled asymmetric supercapacitor delivered an outstanding energy density of 23.25 W h Kg^{-1} with an excellent power density of 9.3 KW Kg^{-1}, which is also used for light-emitting diode (LED). Further, CuO is also recognized as an effective electrode material owing to its elemental abundance. Kaner et al. developed a three-dimensionally ordered CuO framework with bimodal nanopores and nanosized walls like structured materials. The combination of assembled device with an activated carbon electrode exhibited an energy density of 19.7 W h Kg^{-1} and excellent cycling life [111]. Recently, Chang et al demonstrated a yolk-shell-structured $ZnCo_2O_4$ with uniform carbon for an aqueous asymmetric supercapacitor, which exhibited an outstanding cycling stability with a capacitance retention rate of over 95% even after

Bioinspired Nanomaterials for Energy and Environmental Applications   Materials Research Forum LLC
Materials Research Foundations **121** (2022) 141-174   https://doi.org/10.21741/9781644901830-5

9000 cycles and high energy density of 45.9 W h Kg^{-1} with high power density of 700 W Kg^{-1} [112]. As a result, the development of various pseudocapacitive metal oxide heterostructures motivated a broad interest on the synergistic effects of different phases and heterostructures. These approaches essentially provided more active sites and improved the electrical conductivities towards enhancing the electrochemical performances of the asymmetric devices. Accordingly, a hierarchical core-shell-like NiO-Co_3O_4-NiO nano heterostructures were developed by a simple solution-based method in order to increase the interaction abilities of aqueous electrolyte ions diffusing into bulk materials. In their study, the device was assembled as an asymmetric device with NiO-Co_3O_4-NiO with fish thorn-like nanostructure (FTNs) as the positive electrode and porous activated carbon loaded on nickel foam (AC@NF) as the negative. As a result, a coulombic efficiency of 99.7% was achieved and it also demonstrated an outstanding reversibility in the developed supercapacitor device [114-116].

3.7   Transition Metal Sulfides (TMS)

The carbon intergraded materials such as metal oxides/hydroxides and conducting polymers have been explored as a potential electrode material for supercapacitor. Transition metal oxides including MnO_2 and RuO_2 are the most explored candidates for the supercapacitor electrode materials due to their high specific capacitance (theoretical value of ~ 1370 F g^{-1} for MnO_2 and around 1500-2200 F g^{-1} for RuO_2), reversible charge-discharges behavior, better electrical conductivity (10^{-5}-10^{-6} S cm^{-1}). However, the poor rate scan performance of MnO_2, more processing time and expensive costs of the RuO_2 electrode materials severely affected their commercial applications. The main drawback of transition metal oxides is their poor electrical conductivity, which fundamentally decreases the overall performance of the electrochemical device. Alternatively, the transition metal sulfides (TMS) have gained more attention as an electrode material for supercapacitors due to their its exclusive physicochemical properties, low price, abundance in earth, ease synthesis and high theoretical capacities value, which often make them possible as the substitutes for noble metals. The structure profile, and particle size of the TMSs play a vital role in the electrocatalytic applications. As compared to TMO, the TMS based electrode materials exhibit higher conductivity owing to their faradic reactions involved in the charges-discharge processes. Most importantly, the electrochemical activity of TMS is higher that of TMO due to the substitution because of presents of oxygen with sulphur atoms, which are higher ionic conductivity. Further, TMS have high ionic diffusivity because of their higher anionic polarizability. Among the different TMSs studied, the cobalt sulfide (CoS), copper sulfide (CuS), iron sulfide (FeS), and nickel sulfide (NiS) are found to be significant multifunctional semiconductors for supercapacitor electrode applications [117-120].

In general, the TMSs can be classified as two types, (i) layered sulfides and (ii) non-layered sulfides. The layered metals sulfides such as MoS_2, WS_2, SnS_2, VS_2 etc. are composed of three atom layers (S-M-S), which are formed through Vander-Waals forces. There are different methods available to synthesis the metal sulfides such as, hydrothermal/solvothermal, precipitation, electro-deposition, microwave, atomic layer deposition, ball milling, chemical vapour deposition, thermal deposition, in-situ polymerization, exfoliation, and electrochemical synthesis. Among them, the solid-state synthesis and exfoliation of MSs at elevated temperature are some of the most important synthesis methods. The high temperature methods require more reaction time and also expensive and therefore it restricts their usage in commercial/industrial application [121-125]. Recently, Durga et al. fabricated four different metal sulfides grown onto the nickel foam (CoC/NF,CuS/NF, FeS/NF, and NiS/NF) by simple chemical bath deposition method for supercapacitor application. Among the various metal sulfides, FeS/NF exhibited the better specific capacitance (2007.61 F g^{-1} at 2 A g^{-1}) along with excellent cycling stability and good rate capability. The high specific capacitance was achieved in FeS/NF materials due to their large surface area, which provided the large interfacial area between the electrode and the electrolyte [125].

3.8   Bioinspired conducting polymers (CPs) nanostructures

The conducting polymers (CPs) are basically organic polymers, which provide the excellent electricity by conjugated bond. The CPs is widely used in supercapacitors. Recently, the conducting polymers such as polyaniline, polypyrole, polythiophene and poly(3,4-ethylenedioxythiophene) are used as electrode materials in supercapacitors due to their inexpensive, high pseudocapacitance and high conductivity. The size and shape of CPs play vital roles in their electrochemical performance. CPs can be easily prepared by chemical or electrochemical polymerization and can be obtained as gel or bulk powder, films and sponges. Polyaniline (PANI) can be synthesized by oxidative polymerization and used as a positive electrode material in pseudocapacitance. PANI typically delivers monotonically charge/discharge potential window from -0.2 to +0.8 V. On the other hand, polypyrole is widely used as electrode materials due to their enhanced conductivity as compared to the PANI and polythiophene. Polypyrole can be synthesized by potentiodynamic deposition between 0 V and +1.2 V on a stainless steel foil using 0.05 M pyrrole solution and 0.1 M KNO_3. The electrochemical performance of pyrrole based electrodes is found to be dependent on their size, scan rate, porosity and morphology [127-130]. Wang et. al., reported the bioinspired high performance electrochemical devices using conducting polymers based coral materials [54]. Various electrode materials were also prepared by depositing the conducting polymers such as PANI, polythiophene and polypyrrole on the monolithic coral like porous carbon.

It should be noted that the coral structure has peculiar dendritic structure and there are plenty of pores with different size and thus they play different roles in electrochemical performance. For instance, the porous dendritic structures provide a large surface area for the conventional ion exchange process. These hierarchical structures demonstrate good capacitance due to the presence of conducting polymers. The composite electrode showed the excellent specific capacitance up to 1488 F g^{-1} at current density of 1.0 A g^{-1} and it showed energy density of 49.5 Wh kg^{-1}. Hence, these hybrid composite materials would be ideal materials to developed the electrodes for supercapacitor. Moreover, their morphological structure in microscopic images clearly show the skeleton structure and replicated macroporous channels. In the natural coral structure, the pore diameter appears to be skeleton like structures as clearly observed in SEM and optical microscopy as shown in Fig 8. The pore diameter of the macropores channel ensures the diffusion of ions and electrolytes inside the carbon monolith smoothly [54].

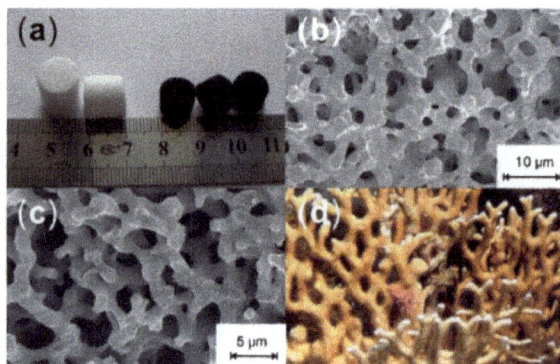

Figure 8. (a) Prepared carbon materials, (b and c) SEM images of silica template and carbon materials (d) optical image of coral. Reproduced permission from ref no [54].

Concluding remarks and future prospects

Bioinspired nanomaterials hold the key to fundamental advances in energy conversion and storage. Nanomaterials can offer unique properties in electrodes and electrolytes based energy devices. The increasing industrializations necessitate the rapid and abundant electrochemical energy production to meet the demands. The instant energy conversion and utilization may not be the comprehensive solution, and therefore it is as equally as important to develop efficient materials towards storing the produced energy. In this direction, supercapacitors have been widely recognized as the promising devices for the energy storage due to their high power density, improved cycling and fast charge-discharge mechanism. Towards developing efficient materials to construct high efficient electrodes for supercapacitors, the emerging bioinspired nanomaterials are widely explored and used. These bioinspired nanomaterials mainly composed of hierarchical structures and constructed as dictated by the natural biological structures. In this context, this book chapter deals with latest progresses in the design and development of nanoelectrolytes and nanoelectrodes using various biological nanostructures such as peptide based nanotubes, peptide modified carbon electrode, conducting polymers modified monolithic carbon materials and bioinspired 2D nanomaterial for supercapacitor applications.

Acknowledgements

The author (Dr. A. Murali) grateful to the Department of Science and Technology (DST), Govt. of India, grant no: DST/INSPIRE/04/2018/001762 for Inspire Faculty scheme.

References

[1] J. Yan, Q. Wang, T. Wei and Z. J. Fan, Recent advances in design and fabrication of electrochemical supercapacitors with high energy densities, Adv. Energy Mater. 4 (2014) 1300816. https://doi.org/10.1002/aenm.201300816

[2] L. M. Dai, D. W. Chang, J. B. Baek and W. Lu, Carbon nanomaterials for advanced energy conversion and storage, Small, 8 (2012)1130-1166. https://doi.org/10.1002/smll.201101594

[3] Y. C. Liu, B. B. Huang, X. X. Lin and Z. L. Xie, Biomass-derived hierarchical porous carbons: boosting the energy density of supercapacitors via an ionothermal approach, J. Mater. Chem. A, 5 (2017) 25090. https://doi.org/10.1039/C7TA90265D

[4] J. P. Holdren, Energy and sustainability, Science 315 (2007) 737. https://doi.org/10.1126/science.1139792

[5] G. Wang, L. Zhang, J. Zhang, A review of electrode materials for electrochemical supercapacitors, Chem. Soc. Rev. 41 (2012) 797-828. https://doi.org/10.1039/C1CS15060J

[6] Y. Huang, J. Liang, Y. Chen, An overview of the applications of graphene-based materials in supercapacitors, Small 8 (2012) 1805-1834. https://doi.org/10.1002/smll.201102635

[7] L. Ji, P. Meduri, V. Agubra, X. Xiao, and M. Alcoutlabi, Graphene-Based Nanocomposites for Energy Storage, Adv. Energy Mater. (2016) 1502159. https://doi.org/10.1002/aenm.201502159

[8] H. Lu, X. S. Zhao, Biomass-derived carbon electrode materials for supercapacitors, Sustainable Energy Fuels 1 (2017) 1265-1281. https://doi.org/10.1039/C7SE00099E

[9] Z. Yu, L. Tetard, L. Zhai and J. Thomas, Supercapacitor electrode materials: nanostructures from 0 to 3 dimensions, Energy & Environmental Science, 8 (2015) 702-730. https://doi.org/10.1039/C4EE03229B

[10]   L. L. Zhang, Y. Gu and X. Zhao, Advanced porous carbon electrodes for electrochemical capacitors, J. Mater. Chem. A, 1(2013) 9395-9408. https://doi.org/10.1039/c3ta11114h

[11]   S. Saha, P. Samanta, N. C. Murmu, T. Kuila, A review on the heterostructure nanomaterials for supercapacitor application, J. Energy Storage 17 (2018) 181-202. https://doi.org/10.1016/j.est.2018.03.006

[12]   H. Zhang, X. Lv, Y. Li, Y. Wang, J. Li, P25-Graphene composite as a high performance photocatalyst, ACS Nano 4 (2010) 380-386. https://doi.org/10.1021/nn901221k

[13]   E. Munch, M.E. Launey, D. H. Alsem, E. Saiz, A. P. Tomsia, R.O. Ritchie Tough, Bio-Inspired Hybrid Materials, Science 322 (2008) 1516. https://doi.org/10.1126/science.1164865

[14]   C. Tamerler, M. Sarikaya, Molecular biomimitics: Utilizing naturels molecular ways in practical engineering, Acta Biomater 3 (2007) 289. https://doi.org/10.1016/j.actbio.2006.10.009

[15]   F. Bonaccorso, L. Colombo, G. H. Yu, M. Stoller, V. Tozzini, A. C. Ferrari, R. S. Ruoff and V. Pellegrini, 2D Materials. Graphene, Related Two-Dimensional Crystals, and Hybrid Systems for Energy Conversion and Storage Science, 347 (2015)1246501. https://doi.org/10.1126/science.1246501

[16]   S. Iijima, Helical microtubules of graphitic carbon, Nature 354 (1991) 56. https://doi.org/10.1038/354056a0

[17]   G. Jorio, Dresselhaus, M. S. Dresselhaus, "Carbon Nanotubes: Advanced topics in synthesis, structure and properties", Springer, New York, (2007). https://doi.org/10.1007/978-3-540-72865-8

[18]   S. Iijima, and T. Ichihashi. Single-shell carbon nanotubes of 1-nm diameter, Nature, 363 (1993) 603. https://doi.org/10.1038/363603a0

[19]   T. Guo, P. Nikolaev, A. Thess, D. T. Colbert, and R. E. Smalley, Catalytic growth of single-walled manotubes by laser vaporization, Chem. Phys. Lett,. 243 (1995) 49. https://doi.org/10.1016/0009-2614(95)00825-O

[20]   P. Nikolaev, M. J. Bronikowski, R. K. Bradley, F. Rohmund, D. T. Colbert, K. A. Smith, Gas-phase catalytic growth of single-walled carbon nanotubes from carbon monoxide, Chem. Phys. Lett., 313, (1999) 91. https://doi.org/10.1016/S0009-2614(99)01029-5

[21]   M. Joseyacaman, M. Mikiyoshida, L. Rendon, and J. G. Santiesteban. Catalytic growth of carbon microtubules with fullerene structure, Appl. Phys. Lett., 62 (1993) 657. https://doi.org/10.1063/1.108857

[22]   J. Li, and Y. Zhang, A simple purification for single-walled carbon nanotubes, Physica E., 28 (2005) 309. https://doi.org/10.1016/j.physe.2005.03.022

[23]   M. F. Yu, B. S. Files, S. Arepalli, and R. S. Ruoff, Tensile Loading of Ropes of Single Wall Carbon Nanotubes and Their Mechanical Properties, Phys. Rev. Lett., 84 (2000) 5552. https://doi.org/10.1103/PhysRevLett.84.5552

[24]   P. M. Ajayan, O. Stephan, C. Colliex, and D. Trauth, Aligned Carbon Nanotube Arrays Formed by Cutting a Polymer Resin-Nanotube Composite, Science, 265(1994) 1212. https://doi.org/10.1126/science.265.5176.1212

[25]   G. L. Hornyak, J. Dutta, H. F. Tibbals, and A. K. Rao, *Introduction to Nanoscience*, CRC Press, Taylor and Francis Group, Boca Raton, (2008). https://doi.org/10.1201/b12835

[26]   P. M. Ajayan, and O. Z. Zhou, *Applications of carbon nanotubes*, In Carbon nanotubes, Springer Berlin Heidelberg, pp. 391(2001). https://doi.org/10.1007/3-540-39947-X_14

[27]   A.Jorio, G. Dresselhaus and M. S. Dresselhaus, *Carbon Nanotubes: Advanced Topics in the Synthesis, Structure, Properties and Applications,* Springer-Verlag, Berlin, Heidelberg, (2008). https://doi.org/10.1007/978-3-540-72865-8

[28]   R. C. Haddon, Carbon Nanotubes Acc. Chem. Res.,35(2002) 997. https://doi.org/10.1021/ar020259h

[29]   P. Singh, S. Campidelli, S. Giordani, D. Bonifazi, A. Biancoa and M. Prato, Organic Functionalisation and Characterisation of Single-Walled Carbon Nanotubes, Chem. Soc. Rev., 38(2009) 2214. https://doi.org/10.1039/b518111a

[30]   W. Ranran, J. Sun, L. Gao, and J. Zhang, Base and Acid Treatment of SWCNT-RNA Transparent Conductive Films, ACS nano., 4 (2010) 4890. https://doi.org/10.1021/nn101208m

[31]   C. Niu, E.K. Sichel, R. Hoch, D. Moy, H. Tennent, High power electrochemical capacitors based on carbon nanotube electrodes, Appl. Phys. Lett. 70 (1997) 1480. https://doi.org/10.1063/1.118568

[32]   J.N. Barisci, G.G. Wallace, R.H. Baughman, Electrochemical Characterization of Single-Walled Carbon Nanotube Electrodes, J. Electrochem. Soc. 147 (2000) 4580. https://doi.org/10.1149/1.1394104

[33]   S. Shiraishi, H. Kurihara, K. Okabe, D. Hulicova, A. Oya, Electric double layer capacitance of multi-walled carbon nanotubes and B-doping effect, Electrochem. Commun. 4(2002) 593. https://doi.org/10.1016/S1388-2481(02)00382-X

[34]   V. Romano, B. Martín-García, S. Bellani, L. Marasco, J. Kumar Panda, R. Oropesa-Nuñez, L. Najafi, A. Esau Del Rio Castillo, M. Prato, E. Mantero, V. Pellegrini, Giovanna D'Angelo, and F. Bonaccorso, Flexible Graphene/Carbon Nanotube Electrochemical Double-Layer Capacitors with Ultrahigh Areal Performance, Chem Plus Chem, *84* (2019) 882–892. https://doi.org/10.1002/cplu.201900235

[35]   J. Gamby, P. L. Taberna, P. Simon, J. F. Fauvarque, M. Chesneau, Studies and characterisations of various activated carbons used for carbon/carbon supercapacitors J. Power Sources 101 (2001) 109–116. https://doi.org/10.1016/S0378-7753(01)00707-8

[36]   L. Permann, M. Lätt, J. Leis, M. Arulepp, Electrical double layer characteristics of nanoporous carbon derived from titanium carbide, Electrochim. Acta 51 (2006) 1274–1281. https://doi.org/10.1016/j.electacta.2005.06.024

[37]   R.Shah, X. Zhang and S. Talapatra, Electrochemical double layer capacitor electrodes using aligned carbon nanotubes grown directly on metals, Nanotechnology 20 (2009) 395202. https://doi.org/10.1088/0957-4484/20/39/395202

[38]   J. Guo, Y. Yu, L. Sun, Z. Zhang, R. Chai, K. Shi, Y. Zhao, Bio-inspired multicomponent carbon nanotube microfibers from microfluidics for supercapacitor, Chem Eng. J. (2020), 125517. https://doi.org/10.1016/j.cej.2020.125517

[39]   (a) G. Xiong, P. He, Z. Lyu, T. Chen, B. Huang, L. Chen & T. S. Fisher, Bioinspired leaves-on-branchlet hybrid carbon nanostructure for supercapacitors, Nat Commun. 9, (2018) 790; https://doi.org/10.1038/s41467-018-03112-3 (b) Y. Zhou, Y. Zhu, B. Xu, and X. Zhang, Chem. Commun. 55 (2019) 4083-4086. https://doi.org/10.1039/C9CC01277J

[40]   T. Chen, L. Dai, Carbon nanomaterials for high-performance supercapacitors. Mater. Today 16 (2013) 272–280. https://doi.org/10.1016/j.mattod.2013.07.002

[41]   H. Zhang, G.Cao, Z. Wang, Y. Yang, Z. Shi, and Z. Gu Growth of manganese oxide nanoflowers on vertically-aligned carbon nanotube arrays for high-rate electrochemical capacitive energy storage. Nano. Lett. 8 (2008) 2664-2668. https://doi.org/10.1021/nl800925j

[42]   Y. Shin-Yi, C. Kuo-Hsin, H. Tien , L.Ying-Feng, S. Li , Y. Wang , J. Wang , C. M. Ma , C. Hu, Design and tailoring of a hierarchical graphene-carbon nanotube architecture for supercapacitors. J. Mater. Chem. 21 (2011) 2374–2380. https://doi.org/10.1039/C0JM03199B

[43]   A. Kumar, M.R. Maschmann, S. L. Hodson, J. Baur, T. S. Fisher, Carbon nanotube arrays decorated with multi-layer graphene-nanopetals enhance mechanical strength and durability. Carbon 84 (2015) 236–245. https://doi.org/10.1016/j.carbon.2014.11.060

[44]   X. Zhao, H. Tian, M. Zhu, K. Tian, J. J. Wang, F. Kang, R. A. Outlaw Carbon nanosheets as the electrode material in supercapacitors. J. Power Sources 194, (2009)1208–1212. https://doi.org/10.1016/j.jpowsour.2009.06.004

[45]   L. Jiang, Z. Ren, S. Chen, Q. Zhang, X. Lu, H. Zhang and G. Wan, Bio-derived three-dimensional hierarchical carbon-graphene-TiO_2 as electrode for supercapacitors. Sci Rep 8 (2018) 4412. https://doi.org/10.1038/s41598-018-22742-7

[46]   J. Xu, Z.Tan, W. Zeng, G. Chen, S. Wu, Y. Zhao, K. Ni, Z. Tao, M. Ikram, H.Ji, Y. Zhu, A Hierarchical Carbon Derived from Sponge-Templated Activation of Graphene Oxide for High-Performance Supercapacitor Electrodes. Adv. Mater. 28, (2016) 5222–5228. https://doi.org/10.1002/adma.201600586

[47]   Y. Sun, R. B. Sills,X.Hu, Z. Wei Seh, X. Xiao, H. Xu, W. Luo, H. Jin, Y. Xin, T. Li, Z. Zhang, J. Zhou, W. Cai, Y. Huang, and Y. Cui, A Bamboo-Inspired Nanostructure Design for Flexible, Foldable, and Twistable Energy Storage Devices. Nano Lett. 15, (2015) 3899–3906. https://doi.org/10.1021/acs.nanolett.5b00738

[48]   A. Langlois, and F. Coeuret, Flow-through and flow-by porous electrodes of nickel foam. I. Material characterization. J Appl. Electrochem. 19, (1989) 43–50. https://doi.org/10.1007/BF01039388

[49]   D. H. Jacob, M. Wasala, J. Richie, J. Barron, A. Winchester, S. Ghosh, C. Yang W. Xu, L. Song, S. Kar and S. Talapatra, High Performance Graphene-Based Electrochemical Double Layer Capacitors Using 1-Butyl-1-methylpyrrolidinium tris (pentafluoroethyl) trifluorophosphate Ionic Liquid as an Electrolyte, Electronics, 7(2018), 229. https://doi.org/10.3390/electronics7100229

[50]   K. Singh, S. Kumar, K. Agarwal, K. Soni, V. Ramana Gedela, and K. Ghosh, Three-dimensional Graphene with MoS_2 Nanohybrid as Potential Energy Storage/Transfer Device, Sci Rep., 7 (2017) 9458. https://doi.org/10.1038/s41598-017-09266-2

[51]   B. Senthilkumar, Z. Khan, S. Park, K. Kim, H. Ko and Y. Kim, Highly porous graphitic carbon and $Ni_2 P_2 O_7$ for a high performance aqueous hybrid supercapacitor. J. Mater. Chem. A. 3 (2015) 21553–21561. https://doi.org/10.1039/C5TA04737D

[52]   Q. Liao, N. Li, S. Jin, G. Yang, C.Wang, All-solid-state symmetric supercapacitor based on $Co_2 P_2 O_7$ nanoparticles on vertically aligned graphene. ACS Nano. 9 (2015) 5310–5317. https://doi.org/10.1021/acsnano.5b00821

[53]   D.D. Nguyen, C. Hsiao, T. Su, P. Hsieh, Y. Chen, Y. Chueh, C. Lee and N. Tai. Bioinspired networks consisting of spongy carbon wrapped by graphene sheath for flexible transparent supercapacitors. Commun Chem 2(2019) 137. https://doi.org/10.1038/s42004-019-0238-9

[54]   Y. Wang, S. Tao, Y. An, S. Wu and C. Meng, Bio-inspired high performance electrochemical supercapacitors based on conducting polymer modified coral-like monolithic carbon J. Mater. Chem. A, 1 (2013)8876. https://doi.org/10.1039/c3ta11348e

[55]   P. Beker and G. Rosenman, Bioinspired nanostructural peptide materials for supercapacitor electrodes J. Mater. Res., 25, (2011) 1661-1666. https://doi.org/10.1557/JMR.2010.0213

[56]   P. Beker, I. Koren, N. Amdursky, E. Gazit, G. Rosenman, Bioinspired peptide nanotubes as supercapacitor electrodes J. Mater. Sci. 45 (2010) 6374–6378. https://doi.org/10.1007/s10853-010-4624-z

[57]   M. R. Ghadiri, J. R. Granja, R. A. Milligan, D. E. McRee, and N. Khazanovich, Self-assembling organic nanotubes based on a cyclic peptide architecture. Nature 372(1994) 709. https://doi.org/10.1038/372709a0

[58]   Y. Boyjoo, Y. Cheng, H. Zhong, H. Tian, J. Pan, V. K. Pareek, S. P. Jiang, J. F. Lamonier, M. Jaroniec and J. Liu, From waste Coco-cola to activated carbons with impressive capabilities for CO_2 adsorption and supercapacitors, Carbon, 116 (2017) 490-499. https://doi.org/10.1016/j.carbon.2017.02.030

[59]   Q. L. Ma, Y. F. Yu, M. Sindoro, A. G. Fane, R. Wang and H. Zhang, Carbon-Based Functional Materials Derived From Waste for Water Remediation and Energy Storage, Adv. Mater., 29 (2017) 1605361. https://doi.org/10.1002/adma.201605361

[60]   Z. Bi, Kong, Y. Cao, G. Sun, F. Su, X. Wei, X. Li, A. Ahmad, L. Xie and C. Chen, Biomass-derived porous carbon materials with different dimensions for supercapacitor electrodes: a review, J. Mater. Chem. A, 7 (2019)16028-16045. https://doi.org/10.1039/C9TA04436A

[61]   W. F. Chen, S. Iyer, S. Iyer, K. Sasaki, C. H. Wang, Y. M. Zhu, J. T. Muckerman and E. Fujita, Biomass-derived electrocatalytic composites for hydrogen evolution, Energy Environ. Sci., 6 (2013)1818-1826. https://doi.org/10.1039/c3ee40596f

[62]   J. Z. Chen, K. L. Fang, Q. Y. Chen, J. L. Xu and C. P. Wong, P. Integrated paper electrodes derived from cotton stalks for high-performance flexible supercapacitors Nano Energy, 53 (2018) 337-344. https://doi.org/10.1016/j.nanoen.2018.08.056

[63]   S. J. Song, F. W. Ma, G. Wu, D. Ma, W. D. Geng and J. F. Wan, Facile self-templating large scale preparation of biomass-derived 3D hierarchical porous carbon for advanced supercapacitors, J. Mater. Chem. A, 3 (2015)18154-18162. https://doi.org/10.1039/C5TA04721H

[64]   J. Wang, P. Nie, B. Ding, S. Y. Dong, X. D. Hao, H. Dou and X. G. Zhang, Biomass derived carbon for energy storage devices, J. Mater. Chem. A, 2017, 5, 2411-2428. https://doi.org/10.1039/C6TA08742F

[65]   M. Liu, K. Zhang, M. Si, H. Wang, L. Chai, Y. Shi, Three-dimensional carbon nanosheets derived from micro-morphologically regulated biomass for ultrahigh-performance supercapacitors, Carbon 153 (2019) 707-716. https://doi.org/10.1016/j.carbon.2019.07.060

[66]   H. Lu, L. Zhuang, R. Gaddam, X. Sun, C. Xiao, T. Duignan, Z. Zhu, X. Zhao, Microcrystalline cellulose-derived porous carbons with defective sites for electrochemical applications, J. Mater. Chem. 7 (2019) 22579-22587. https://doi.org/10.1039/C9TA05891E

[67]   S. Liu, Y. Zhao, B. Zhang, H. Xia, J. Zhou, W. Xie, H. Li, Nano-micro carbon spheres anchored on porous carbon derived from dual-biomass as high rate performance supercapacitor electrodes, J. Power Sources 381 (2018) 116-126. https://doi.org/10.1016/j.jpowsour.2018.02.014

[68]   J. Mi, X-R. Wang, R-J.Fan, W-H.Qu, WC.Li, Coconut-shell-based porous carbons with a tunable micro/mesopore ratio for high-performance supercapacitors, Energy Fuels 26 (2012) 5321-9. https://doi.org/10.1021/ef3009234

[69]   Y. Li, G. Wang, T. Wei, Z. Fan, P. Yan, Nitrogen and sulfur co-doped porous carbon nanosheets derived from willow catkin for supercapacitors, Nano Energy 19 (2016) 165-75. https://doi.org/10.1016/j.nanoen.2015.10.038

[70]   W. Tian, Q. Gao, Y. Tan, K. Yang, L. Zhu, C. Yang, H. Zhang, Bio-inspired beehive-like hierarchical nanoporous carbon derived from bamboo-based industrial by-product as a high performance supercapacitor electrode material, J. Mater. Chem. A 3 (2015) 5656-5664. https://doi.org/10.1039/C4TA06620K

[71]   Y. Lv, L. Gan, M. Liu, W. Xiong, Z. Xu, D. Zhu, D.S. Wright, A self-template synthesis of hierarchical porous carbon foams based on banana peel for supercapacitor electrodes, J. Power Sources 209 (2012) 152-157. https://doi.org/10.1016/j.jpowsour.2012.02.089

[72]   L. Zhang, T. You, T. Zhou, X. Zhou, F. Xu, Interconnected hierarchical porous carbon from lignin-derived byproducts of bioethanol production for ultra-high performance supercapacitors, ACS Appl. Mater. Inter. 8 (2016) 13918-13925. https://doi.org/10.1021/acsami.6b02774

[73]   C. Zhao, Y. Huang, C. Zhao, S. Shao, Z. Zhu, Rose-derived 3D carbon nanosheets for high cyclability and extended voltage supercapacitors, Electrochim. Acta 291 (2018) 287-296. https://doi.org/10.1016/j.electacta.2018.09.136

[74]   F. H-h, L. Chen, H. Gao, X. Yu, J. Hou, G. Wang, F. Yu, H. Li, C. Fan, Y-l.Shi, X. Guo, Walnut Shell-derived hierarchical porous carbon with high performances for electrocatalytic hydrogen evolution and symmetry supercapacitors, Int. J. Hydrogen Energ.45 (2020) 443-51. https://doi.org/10.1016/j.ijhydene.2019.10.159

[75]   Q. Xiong, Q. Bai, C. Li, D. Li, X. Miao, Y. Shen, H. Uyama, Nitrogen-doped hierarchical porous carbons from used cigarette filters for supercapacitors, J. Taiwan. Inst. Chem. E. 95 (2019) 315-323. https://doi.org/10.1016/j.jtice.2018.07.019

[76]   M-S. Wu, L-J. Lyu, J-H. Syu, Copper and nickel hexacyanoferrate nanostructures with graphene-coated stainless steel sheets for electrochemical supercapacitors, Journal of Power Sources 297 (2015) 75-82. https://doi.org/10.1016/j.jpowsour.2015.07.101

[77]   J. Chen, K. Huang, S. Liu, Insoluble metal hexacyanoferrates as supercapacitor electrodes, Electrochem. Commun. 10 (2008) 1851–1855. https://doi.org/10.1016/j.elecom.2008.07.046

[78]   F. Zhao, Y. Wang, X. Xu, Y. Liu, R. Song, G. Lu, Y. Li, ACSAppl. Mater. Interfaces 6, 11007–11012 (2014). https://doi.org/10.1021/am503375h

[79]   A. Safavi, S. H. Kazemi, H. Kazemi, Electrochemically deposited hybrid nickel-cobalt hexacyanoferrate nanostructures for electrochemical applications. Electrochem Acta 56 (2011)9191. https://doi.org/10.1016/j.electacta.2011.07.122

[80]   Y. Wang, Y. Yang, X. Zhang, C. Liu, X. Hao, One-step electrodeposition of polyaniline/nickel hexacyanoferrate/sulfonated carbon nanotubes interconnected composite films for supercapacitor, J. Solid State Electrochem.19(2015) 3157-3168. https://doi.org/10.1007/s10008-015-2934-4

[81]   R.S. Babu, A.L.F. de Barros, M.A. Maier, D.M.Sampaio, J. Balamurugan, J.H. Lee, Novel polyaniline/manganese hexacyanoferrate nanoparticles on carbon fiber as binder-free electrode for flexible supercapacitors, Compos. Part B Eng. 143 (2018) 141-147. https://doi.org/10.1016/j.compositesb.2018.02.007

[82]   M. A. Maier, R.S. Babu, D.M. Sampaio, A.L.F. de Barros, Binder-free polyaniline interconnected metal hexacyanoferrates nanocomposites (Metal = Ni, Co) on carbon fibers for flexible supercapacitors, J. Mater. Sci. Mater. Electron. 28 (2017) 17405-17413. https://doi.org/10.1007/s10854-017-7674-z

[83]   G. Wang, L. Zhang and J. Zhang, A Review of Electrode Materials for Electrochemical Supercapacitors, Chem. Soc. Rev., 41 (2012)797–828. https://doi.org/10.1039/C1CS15060J

[84]   G. Yu, L. Hu, N. Liu, H. Wang, M. Vosguerichian, Y. Yang, Y. Cui and Z. Bao, Enhancing the Supercapacitor Performance of Graphene/MnO2 Nanostructured Electrodes by Conductive Wrapping, Nano Lett., 11 (2011) 4438–4442. https://doi.org/10.1021/nl2026635

[85]   Q. Lu, J. Chen and J. Xiao, Nanostructured Electrodes for High-Performance Pseudocapacitors, Angew. Chem., Int. Ed., 52 (2013) 1882–1889. https://doi.org/10.1002/anie.201203201

[86]   Z. Yu, L. Tetard, L. Zhai and J. Thomas, Supercapacitor electrode materials: nanostructures from 0 to 3 dimensions, Energy Environ. Sci., 8 (2015) 702–730. https://doi.org/10.1039/C4EE03229B

[87]   X. Lu, C. Wang, F. Favier and N. Pinna, Electrospun Nanomaterials for Supercapacitor Electrodes: Designed Architectures and Electrochemical Performance, Adv. Energy Mater., 7 (2017)1601301. https://doi.org/10.1002/aenm.201601301

[88]   E. Johan ten Elshof, H. Yuan, P. Gonzalez Rodriguez, Two-dimensional metal oxide and metal hydroxide nanosheets: synthesis, controlled assembly and applications

in energy conversion and storage. Adv Energy Mater 6 (2016)1600355.
https://doi.org/10.1002/aenm.201600355

[89]   H. Xia, Y. Shirley Meng, G. Yuan, C. Cui, and L. Luc, A symmetric RuO2/RuO2
supercapacitor operating at 1.6 V by using a neutral aqueous electrolyte. Electrochem
Solid-State Lett 15 (2012)A60–A63. https://doi.org/10.1149/2.023204esl

[90]   A.Bello, O.O. Fashedemi, J.N. Lekitima, M.Fabiane, D. Dodoo-Arhin, K. I.
Ozoemena, Y. Gogotsi, A. T. Charlie Johnson, and N. Manyala High-performance
symmetric electrochemical capacitor based on graphene foam and nanostructured
manganese oxide. AIP Adv 3 (2013)82118. https://doi.org/10.1063/1.4819270

[91]   K. Makgopa, P.M. Ejikeme, C.J. Jafta et al, A high-rate aqueous symmetric
pseudocapacitor based on highly graphitized onion-like carbon/birnessite-type
manganese oxide nanohybrids. J Mater Chem A 3 (2015)3480–3490.
https://doi.org/10.1039/C4TA06715K

[92]   J. Yan, E. Khoo, A. Sumboja, P. S. Lee, Facile coating of manganese oxide on tin
oxide nanowires with high-performance capacitive behavior. ACS Nano 4 (2010) 4247–
4255. https://doi.org/10.1021/nn100592d

[93]   D. Wei, M.R.J Scherer, C. Bower et al., A nanostructured electrochromic
supercapacitor. Nano Lett 12 (2012)1857–1862. https://doi.org/10.1021/nl2042112

[94]   N-L.Wu, S-Y. Wang, C-Y. Han et al., Electrochemical capacitor of magnetite in
aqueous electrolytes. J Power Sources 113(2003)173–178. https://doi.org/10.1016/S0378-
7753(02)00482-2

[95]   T.P. Gujar, V.R. Shinde, C.D. Lokhande, S-H. Han, Electrosynthesis of Bi2O3
thin films and their use in electrochemical supercapacitors. J Power Sources
161(2006)1479–1485. https://doi.org/10.1016/j.jpowsour.2006.05.036

[96]   S.K. Deb, Opportunities and challenges in science and technology of WO3 for
electrochromics and related applications. Sol Energy Mater Sol Cells 92 (2008) 245–258.
https://doi.org/10.1016/j.solmat.2007.01.026

[97]   G. Zhou, D-W. Wang, P-X. Hou et al., A nanosized Fe2O3 decorated single-
walled carbon nanotube membrane as a high-performance flexible anode for lithium ion
batteries. J Mater Chem 22(2012)17942. https://doi.org/10.1039/c2jm32893c

[98]   S. Trasatti and G. Buzzanca, J. Electroanal. Chem., 29 (1971)l–5.
https://doi.org/10.1016/S0022-0728(71)80111-0

[99]   S. Kong, K. Cheng, T. Ouyang, Y. Gao, K. Ye, G. Wang and D. Cao, Facile
electrodepositing processed of RuO2-graphene nanosheets-CNT composites as a binder-

free electrode for electrochemical supercapacitors, Electrochim. Acta, 246 (2017) 433–442. https://doi.org/10.1016/j.electacta.2017.06.019

[100] W. Zhang, H. Lin, H. Kong, H. Lu, Z. Yang and T. Liu, High energy density PbO2/activated carbon asymmetric electrochemical capacitor based on lead dioxide electrode with three-dimensional porous titanium substrate, Int. J. Hydrogen Energy, 39 (2014)17153–17161. https://doi.org/10.1016/j.ijhydene.2014.08.039

[101] S. K. Meher and G. Ranga Rao, Ultralayered Co3O4 for High-Performance Supercapacitor Applications, . Phys. Chem. C 115, (2011)15646–15654. https://doi.org/10.1021/jp201200e

[102] T. Xiong, T. L. Tan, L. Lu, W. S. V. Lee and J. Xue, Adv. Energy Mater., 8 (2018) 1702630. https://doi.org/10.1002/aenm.201702630

[103] Y. Liu, L. Guo, X. Teng, J. Wang, T. Hao, X. He and Z. Chen, High-performance 2.5 V flexible aqueous asymmetric supercapacitors based on K+/Na+-inserted MnO2 nanosheets, Electrochim. Acta, 300 (2019)9–17. https://doi.org/10.1016/j.electacta.2019.01.087

[104] T. Liu, L. Zhang, W. You and J. Yu, Core–Shell Nitrogen-Doped Carbon Hollow Spheres/Co3O4 Nanosheets as Advanced Electrode for High-Performance Supercapacitor, Small, 14 (2018)1702407. https://doi.org/10.1002/smll.201702407

[105] G. Saeed, S. Kumar, N. H. Kim and J. H. Lee, Fabrication of 3D graphene-CNTs/α-MoO3 hybrid film as an advance electrode material for asymmetric supercapacitor with excellent energy density and cycling life, Chem. Eng. J., 352 (2018) 268–276. https://doi.org/10.1016/j.cej.2018.07.026

[106] F. Zheng, C. Xi, J. Xu, Y. Yu, W. Yang, P. Hu, Y. Li, Q. Zhen, S. Bashir and J. L. Liu, Facile preparation of WO3 nano-fibers with super large aspect ratio for high performance supercapacitor, J. Alloys Compd., 772 (2019) 933–942. https://doi.org/10.1016/j.jallcom.2018.09.085

[107] X. Yuan, B. Chen, X. Wu, J. Mo, Z. Liu, Z. Hu, z. Liu, c. Zhou, h. Yang and Y. Wu, An Aqueous Asymmetric Supercapacitor Based on Activated Carbon and Tungsten Trioxide Nanowire Electrodes, Chin. J. Chem., 35(2017) 61–66. https://doi.org/10.1002/cjoc.201600212

[108] S.Jiao, T. Li,C. Xiong, C. Tang, A. Dang, H. Li, and T. Zhao, A Facile Method of Preparing the Asymmetric Supercapacitor with Two Electrodes Assembled on a Sheet of Filter Paper, Nanomaterials (Basel). 9 (2019)1338. https://doi.org/10.3390/nano9091338

[109]　X. Hong, S. Li, R. Wang and J. Fu, Hierarchical SnO2 nanoclusters wrapped functionalized carbonized cotton cloth for symmetrical supercapacitor, J. Alloys Compd., 775(2019) 15–21. https://doi.org/10.1016/j.jallcom.2018.10.099

[110]　D. Wei, M.R.J. Scherer, C.Bower et al A nanostructured electrochromic supercapacitor. Nano Lett 12(2012)1857–1862. https://doi.org/10.1021/nl2042112

[111]　S. E. Moosavifard, M. F. El-Kady, M. S. Rahmanifar, R. B. Kaner and M. F. Mousavi, Designing 3D Highly Ordered Nanoporous CuO Electrodes for High-Performance Asymmetric Supercapacitors, ACS Appl. Mater. Interfaces, 7 (2015) 4851–4860. https://doi.org/10.1021/am508816t

[112]　J. Chang, W. Lee, R.S.Mane et al Morphology-dependent electrochemical supercapacitor properties of indium oxide. Electrochem Solid-State Lett 11 (2008) A9. https://doi.org/10.1149/1.2805996

[113]　V. S. Kumbhar and D. H. Kim, Hierarchical coating of MnO2 nanosheets on ZnCo2O4 nanoflakes for enhanced electrochemical performance of asymmetric supercapacitors, Electrochim. Acta, 271 (2018) 284–296. https://doi.org/10.1016/j.electacta.2018.03.147

[114]　C. Chen, S. Wang, X. Luo, W. J. Gao, G. J. Huang, Y. Zeng and Z. H. Zhu, Reduced ZnCo2O4@NiMoO4·H2O heterostructure electrodes with modulating oxygen vacancies for enhanced aqueous asymmetric supercapacitors, J. Power Sources, 409 (2019)112–122. https://doi.org/10.1016/j.jpowsour.2018.10.066

[115]　S. C. Sekhar, G. Nagaraju and J. S. Yu, High-performance pouch-type hybrid supercapacitor based on hierarchical NiO-Co3O4-NiO composite nanoarchitectures as an advanced electrode material, Nano Energy, 48(2018) 81–92. https://doi.org/10.1016/j.nanoen.2018.03.037

[116]　Z. Li, J. Han, L. Fan and R. Guo, Template-free synthesis of Ni7S6 hollow spheres with mesoporous shells for high performance supercapacitors, Cryst Eng Comm, 17(2015) 1952-1958. https://doi.org/10.1039/C4CE02548B

[117]　N. Li, T. Lv, Y. Yao, H. Li, K. Liu and T. Chen, Compact graphene/MoS2 composite films for highly flexible and stretchable all-solid-state supercapacitors, J. Mater. Chem. A, 5(2017) 3267-3273. https://doi.org/10.1039/C6TA10165H

[118]　S. Sahoo, R. Mondal, D. J. Late and C. S. Rout, Electrodeposited Nickel Cobalt Manganese based mixed sulfide nanosheets for high performance supercapacitor application, Microporous Mesoporous Mater., 244(2017) 101-108. https://doi.org/10.1016/j.micromeso.2017.02.043

[119] R. Gao, Q. Zhang, F. Soyekwo, C. Lin, R. Lv, Y. Qu, M. Chen, A. Zhu and Q. Liu, Novel amorphous nickel sulfide@CoS double-shelled polyhedral nanocages for supercapacitor electrode materials with superior electrochemical properties, Electrochem. Acta, 237 (2017) 94-104. https://doi.org/10.1016/j.electacta.2017.03.214

[120] S. Chandrasekaran, L. Yao, L. Deng, C.Bowen, Y. Zhang, S. Chen, Z. Lin, F. Peng, P. Zhang, Recent advances in metal sulfides: from controlled fabrication to electrocatalytic, photocatalytic and photoelectrochemical water splitting and beyond. Chem Soc Rev, 48 (2019)4178-4280. https://doi.org/10.1039/C8CS00664D

[121] X. Rui, H. Tan, Q.Yan, Nanostructured metal sulfides for energy storage. Nanoscale, 6 (2014)9889-9924. https://doi.org/10.1039/C4NR03057E

[122] A. Borenstein, O. Hanna, R. Attias, S. Luski, T. Brousse, D. Aurbach, Carbon-based composite materials for supercapacitor electrodes: a review. J Mater Chem A 5 (2017)12653–12672. https://doi.org/10.1039/C7TA00863E

[123] Y. Shao, F. M. El-Kady, J. Sun, Y. Li, Q. Zhang, M. Zhu, H. Wang, B. Dunn, R. B. Kaner, Design and mechanisms of asymmetric supercapacitors. Chem Rev 118 (2018) 9233–9280. https://doi.org/10.1021/acs.chemrev.8b00252

[124] S. Z. Iro, C. Subramani, S.S. Dash, A brief review on electrode materials for supercapacitor, Int J Electrochem Sci 11 (2016)10628-10643. https://doi.org/10.20964/2016.12.50

[125] K. Durga Ikkurthi, S. Srinivasa Rao, M. Jagadeesh, Araveeti Eswar Reddy, Tarugu Anitha, Hee-Je Kim, Synthesis of nanostructured metal sulfides via a hydrothermal method and their use as an electrode material for supercapacitors, New J.Chem. 42 (2018) 19183-19192. https://doi.org/10.1039/C8NJ04358B

[126] K.S. Ryu, K. M. Kim, N.-G. Park, Y. J. Park, and S. H. Chang. Symmetric redox supercapacitor with conductingpolyaniline electrodes. J. Power Sources 103 (2002)305–309. https://doi.org/10.1016/S0378-7753(01)00862-X

[127] A.Rudge, I. Raistrick, S. Gottesfeld, and J. P. Ferraris. A study of the electrochemical properties of conducting polymers for application in electrochemical capacitors. Electrochim. Acta 39 (1994)273–287. https://doi.org/10.1016/0013-4686(94)80063-4

[128] Burke, R&D considerations for the performance and application of electrochemical capacitors, Electrochim. Acta53 (2007)1083–1091. https://doi.org/10.1016/j.electacta.2007.01.011

Bioinspired Nanomaterials for Energy and Environmental Applications   Materials Research Forum LLC
Materials Research Foundations **121** (2022) 141-174   https://doi.org/10.21741/9781644901830-5

[129] G. L. Wang, L. Zhang, and J. Zhang. A review of electrode materials for electrochemical supercapacitors. Chem. Soc. Rev.41 (2012)797–828. https://doi.org/10.1039/C1CS15060J

[130] E. Frackowiak, V. Khomenko, K. Jurewicz, K. Lota, and F. Béguin. Supercapacitors based on conductingpolymers/nanotubes composites. J. Power Sources153 (2006)413–418. https://doi.org/10.1016/j.jpowsour.2005.05.030

Bioinspired Nanomaterials for Energy and Environmental Applications   Materials Research Forum LLC
Materials Research Foundations **121** (2022) 175-210   https://doi.org/10.21741/9781644901830-6

Chapter 6

Bio-Mediated Synthesis of Nanomaterials for Dye-Sensitized Solar Cells

G. Murugadoss*[1], T.S. Shyju[1], P. Kuppusami[1]

[1]Centre for Nanoscience and Nanotechnology, Sathyabama Institute of Science and Technology, Chennai -600 119, Tamilnadu, India

* murugadoss_g@yahoo.com

Abstract

Preparation of nanomaterials using bio-mediated techniques is a novel and environmentally friendly method. It has been well demonstrated that the environment friendly green synthesis method is more adoptable for synthesis of various type of nanoparticles including metal oxide and metal sulphide-based compounds, nanocomposites, core-shell and quantum dots (QDs) for several applications. This chapter mainly discusses on the bio-mediated synthesis of the nanoparticles for dye-sensitized solar cells (DSSCs) application. In addition, this chapter briefly discusses about synthesis of the nanomaterials using various plant extracts and then how it can be used as photoanode, photocathode, or sensitizer in DSSCs. It also highlights how the nanoparticles influence in improving the photovoltaic performance.

Keywords

Green Synthesis, Dye-Sensitized Solar Cells (DSSCs), Quantum Dots, Electrolyte, Efficiency

Contents

Bio-Mediated Synthesis of Nanomaterials for Dye-Sensitized Solar Cells.175

1. Introduction..176

2. Bio-mediated synthesis of nanoparticles 178

3. Bio-mediated synthesis of nanoparticles for diverse applications 183

4.   Overview of dye sensitized solar cells **183**

**5.   Bio-synthesized nanoparticles for dye-sensitized
solar cell application** **190**

**6.   Bio-mediated approach for the synthesis of metal nanoparticles for
DSSCs application** **190**

    6.1   Ag Nanoparticles **192**

    6.2   Au nanoparticles **193**

**7.   Bio-mediated synthesised metaloxide nanoparticles for DSSCs
application** **195**

8. Bio-synthesized of Core-shell/QDs on DSSCs Application **200**

Summary and conclusion **201**

Reference **203**

1.   Introduction

The development of clean and sustainable energy is of great demand as it has to replace fossil fuel (~85%) and to take care of environmental issues. Wind and solar energy are low-cost and abundant resources in nature which are consistently distributed and also can eliminate the environmental issues. Solar light energy is a potential source to meet the primary energy demands via solar cells. Solar cell is a device, which converts the sun light into electricity. The photovoltaic effect was first invented by French physicist Edmond Becquerel in 19th century (1839). The solar cells are classified as first generation-silicon solar cells, second generation thin film solar cells and third generation solar cells including dye sensitized solar cells (DSSCs), quantum dot solar cells, polymer solar cells, and organic-inorganic hybrid perovskite solar cells.

Development of new novel nanomaterials are more economical, flexible, environmentally friendly and low fabrication cost for optoelectronic devices, such as light emitting diodes, solar cells, supercapacitor and batteries. DSSC has drawn considerable attention worldwide on sustainable and renewable energies and it was first reported by O' Regan and Gratzel in 1991 [1]. Typically, numerous semiconductor metal oxides with various morphologies were applied to serve as a top layer of dense layer, bilayer, scattered layer photo electrodes as shown in Figure 1. There are some important reports which discusses on, metal oxides such as ZnO, TiO_2, SnO_2, CeO_2 and porous Nb_2O_5 semiconductor materials for photoelectrodes. There is as yet a significant concern to find some other option and stable

Bioinspired Nanomaterials for Energy and Environmental Applications   Materials Research Forum LLC
Materials Research Foundations **121** (2022) 175-210   https://doi.org/10.21741/9781644901830-6

photoanodes to improve the performance of DSSC. Numerous research groups working on the morphological depending function of the metal oxide photoanodes to improve the optical absorption of dye molecules, highly conducting one-dimensional (1D) nanowires, nanorods, nanotubes and scattering materials for high photo conversion efficiency (PEC). Moreover, recently, research report on role of single crystalline nanostructures in improving the charge collection, electron transport, with respect to its polycrystalline materials leading to reduced charge recombination in DSSC. In addition to that, they are developing other oxides to enhance the optoelectronic properties, mainly using ZnO, SnO_2, ZrO_2, and CeO_2 because they have attracted tremendous attention in light absorbing properties with wide bandgap, good optical transmittance, high refractive index, as well as strong adhesion and high stability (chemical and thermal).

One of the key components for DSSC is the TiO_2 anode with a high surface area necessary to stack a lot of adsorbed dye and accomplish a high photocurrent. Subsequently, cube shaped TiO_2 and CeO_2 nanoparticles show a strong light-scattering ability for DSSCs because of their high refractive index for visible light and mirror-like facets.

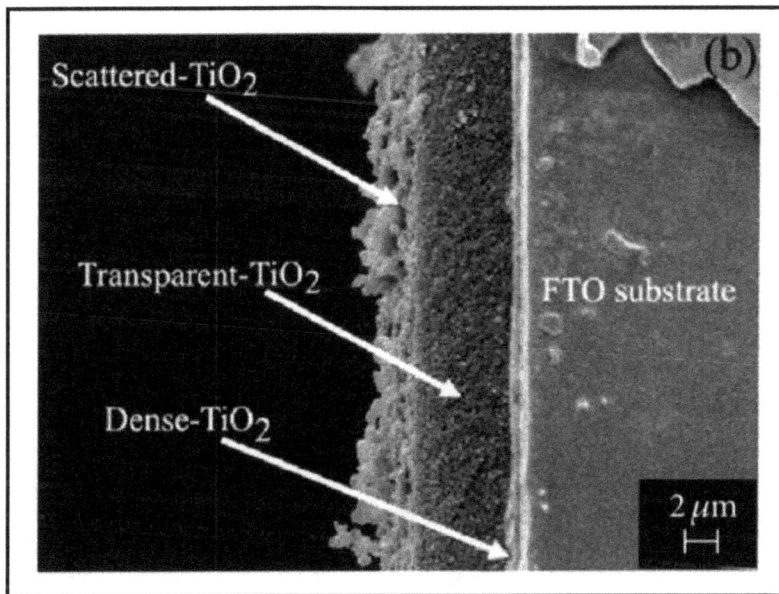

Figure 1 shows the layered structure of DSSC photo electrodes (Copyright from [2])

Researchers have used compact TiO_2 thin film deposited via different facile methods (physical and chemical) to enhance the photovoltaic performance. Now a days, the development and self-assembly of TiO_2 nanostructures with improved light impact and electron transport property are rising as one of the best methodologies for fabricating high efficiency DSSCs. Recently Nobuya Sakai et al [3] reported that ZnO is a promising candidate for photoelectrodes in DSSCs due to cheap semiconductor, porous nature and good electron mobility. Nanocrystalline ZnO can be easily formed via wet chemical synthesis routes, quickly controlled the crystal structure, morphology, particle size shape via varying the temperature duration and composition of the starting precursor. In DSSC, while using ZnO as a photo-electrode one major drawback is corrosion due to acids and it will absorb low amounts of dyes especially for metal organic ruthenium bipyridyl complex dyes. To overcome these issues, researchers reported that immersion of the porous ZnO layer into a $TiCl_4$ treatment at a low temperature to coat an ultrathin layer of TiO_2, for forming core-shell structure of ZnO/TiO_2 structure. To increase the short circuit current (Jsc), open circuit voltage (Voc), and FF with the use of a metalorganic Ru dye (N719) as the sensitizer and details are explained in the following section.

To improve the conversion efficiency of the DSSC, researchers have been working towards development nanomaterials via simple and environmentally friendly methods. Recently many researchers are reporting that green synthesis-based metal and metal oxide nanomaterial's (Au, Ag, Cu, Ni, ZnO, TiO_2, Fe_2O_3, Fe_3O_4, CeO_2, NiO, CuO, and SnO_2) using plants, algae, and microbes as a source of precursor, it has emerged as a green and safe method.

This chapter mainly focuses on the role of nanomaterial in DSSC and synthesis of bio mediated metal oxides for energy conversions. Also, it presents the overview of progress made in developing photoanode and sensitizers in solar cell and to solve the critical issues and also in identifying the key opportunities in the dye-solar business & technology.

2.   Bio-mediated synthesis of nanoparticles

Different techniques were established to synthesis nanoparticles, and the most well-known techniques are solid-state based physical method and chemical method. The detailed description of nanoparticles synthesis is demonstrated in the flow chart (Figure 2). In physical method we need sophisticated techniques in order to achieve the good nanoparticles with different morphology. The most important parameters are good vacuum and high temperature during the preparation process. In chemical method, chemical reaction will take place and we need to understand the chemical kinetics of the starting precursors. In chemical method, we need to control environment to perform the reaction, otherwise the kinetics will differ depending upon the environment and humidity. The main advantage of the chemical method is we can control the morphology of the final product via changing the starting precursors, reducing agents, capping agents, temperature and pressure of the synthesis solution. Also, we can study structure and morphological behaviour via external heat treatment. Recently researchers are concentrating on biological synthesis to develop metal, metal nanoparticles, such as oxides, nitrides and chalcogenides etc., The biological synthesis is a part of chemical method, and in bio synthesis researchers are using extracts from different parts of plant, fungi, viruses, bacteria, macro and microalgae etc.., as a reducing agent to reduce metal ions whichever extracellular or intracellular. In 1999, Gardea-Torresdey et al [4] demonstrated the first synthesis of nanomaterials from biomolecules extracted from plant. The chemical method is an inexpensive and high yield technique compared to physical methods. It's a very easy method for the incorporation of the metals into the nanoparticles or develops composites materials to tune the physical properties such as dispersity, structure, optical, morphology, electrical, mechanical etc.

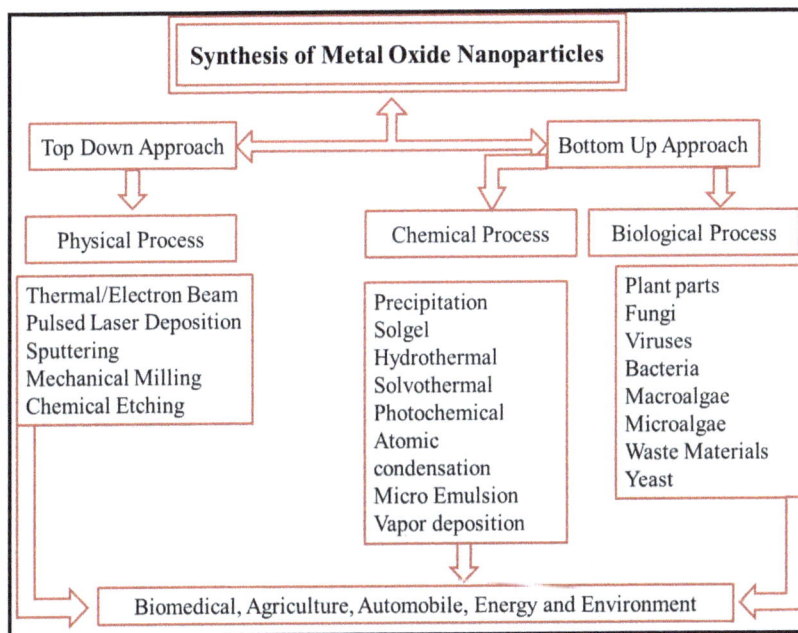

Figure 2. Synthesis processes of metal oxide nanoparticles.

In wet chemistry, precipitation, sol–gel, and solvothermal methods are used as a compelling route for preparation in large-scale. Bio-mediated synthesis method is used for huge production of NPs with biocompatibility, versatility, non-poisonousness, reproducibility and clinical pertinence. In 2016 Slavica Stankic et al reported that, biocompatibility is one of the most significant necessities for nanomaterial used in the field of nano-medicine and molecular research and nanotechnology [5].

The above said method is a good alternative route to prepare metal oxides for biological applications. The detailed description of the plant extract mediated bio synthesis of metal and metal nanoparticle are presented in Table 1. The biological synthesis is simple one-pot synthesis, cost effective, quick and environmentally friendly process for largescale preparation of metal and metal oxide nanoparticles.

Researchers have reported that, [33] plant extracts is much beneficial and they undergo direct reaction in comparison with microorganism. Also, the nanoparticle formation rate is comparatively higher than that of the microorganisms. It does not need any intricate techniques, for example, culture preparation, isolation and culture maintain. Shah et al. [34] reported that, bio mediated synthesis is a low-cost and simple to scale up for the production of huge amounts of nanoparticles. In addition to that, water-soluble biomolecules from plant or microbial biomass play dual roles during the bio mediated synthesis as they can act as protecting and reducing agents. Several functional groups (i.e., hydroxyl, thiol, imidazole, carboxyl, amino) play reductive roles in the formation of metal and metal oxide nanoparticles, because they have strong interaction between biomolecules and metal or metal oxides leading to excellent stability. Proteins are giving various sites (functional groups) to bind an assortment of metal particles. On expansion of metal particles into an aqueous protein solution, the protein molecules quickly bind and catch metal particles. The interaction between metal ions and functional groups can prompt conformational changes of proteins in the solution. Thus, the hydrophobic residues of proteins moreover, go through a self-assembly process under the influence of hydrophobic interactions. The structure and morphology of resulting nanoparticle is therefore reliant on protein/metal particles proportion, temperature, amount of reducing agent, and so on

This chapter mainly focused on the biological synthesis of metal nanoparticles including gold, platinum, silver, palladium, copper, cobalt, cadmium and metal oxides such as titanium dioxide, zinc oxide and so on. In recent literatures, distinctive bacterial strains, for example, escherichia coli, bacillus subtilis, bacillus cereus, pseudo-monas aeruginosa, alter-omonas, bacillus megaterium, klebsiella pneumoniae, and ochrobactrum have been widely utilized for nanoparticles union. Bio mediated synthesis of ZnO nanoparticles using Glycosmis pentaphylla leaf extract and Euphorbia heteradena Jaub root mediated synthesis of TiO_2 nanoparticle have likewise been assessed on the physicochemical properties.

Table 1 gives description of the plant extract mediated bio-synthesis of metal and metal nanoparticles.

S. No.	Plant	Nanoparticle	Shape	Size (nm)	Refs.
1	Sesbania	Au	Spherical	6-20	[6]
2	Alfalfa Sprouts	Ag	Icosahedral Structure	4	[7]
3	Alfalfa plants	Au		10	[8]
4	Capsicum annuum	Ag	Spherical		[9]
5	Aloe vera	Ag/Au	Nanotriangles and Nanoparticles		[10]
6	Populus deltoides	Au	Spherical	20-40	[11]
7	Morinda. Citrifolia leaves	TiO_2	quasi-spherical	10	[12]
8	Cynodon Dactylon	TiO_2	Hexagonal	13-34	[13]
9	Jatropha curcas	TiO_2	Spherical	10-100	[14]
10	Azadirachta indica	ZnO	Hexagonal wurtzite	10-25	[15]
11	Lycopersiconesculentum	ZnO	Spherical shape	70-100	[16]
12	Moringa oleifera	CeO_2	Spherical	45	[17]
13	Honey	CeO_2	Uniform shape	23	[18]
14	Saraca indica	SnO_2	Spherical	2-4	[19]
15	Carica papaya	CuO	Rod	140	[20]
16	Punica granatum peels	Cu			[21]
17	Abutilon indicum	CuO	Hexagonal wurtzite	16	[22]
18	Eucalyptus globulus	NiO	Pleomorphism	20	[23]
19	Citrus sinensis	Fe_2O_3	Spherical	60	[24]
20	Hibiscus Sabdariffa	CdO	Cuboid shape	113	[25]
21	Pelargonium graveolens	Ag	Spherical	16-40	[26]
22	Agathosma betulina	ZnO	quasi-spherical	15	[27]
23	Amorphophallus konjac	ZnO	Rice shaped		[28]
24	Garlic Vine	Fe_2O_3	Hexagonal	16	[29]
25	Filicium decipiens leaf	Pd	Spherical	10-25	[30]
26	Euphorbia Jatropa	ZnO	Hexagonal shape		[31]
27	Coleus forskohlii root	Ag/Au	Spherical and Rod	5-15 & 5-18	[32]

Das et al. [35] synthesized, bio mediated Ag nanoparticles via Bacillus cereus extracted from heavy metal polluted soil and also, they showed surface plasmon resonance properties. These properties are very important in DSSC due to its scattering effects [36]. Mahbubur Rahman et al. 2016 [37] showed the localized surface plasmon resonance effect of nickle (Ni) nanoparticles on the activity of DSSCs for classical I^-/I_3^- electrolyte. Researchers are more focusing on self-assembled peptide nanomaterial as it can be used for light-harvesting systems in solar cells.

Kulkarni et al. [38] have prepared Ag nanoparticles using deinococcus radiodurans with silver chloride precursor. Similarly, researchers are working towards the development of various nanomaterials namely Au, Pd, Pt etc., for various applications. In particular, Ag nanocrystallites are very interesting in the modern research society because of their unique properties, there are broaden ranges of extensive applications in medical industry, cosmetics, catalysis, bioengineering, food packaging, electrochemistry etc., As compared to their bulk materials, nanomaterials exhibit different catalytic activities. Sundrarajan et al. (2017) [12] prepared TiO_2 nanoparticles utilizing leaf extract of Morinda citrifolia (M. citrifolia) by the novel hydrothermal technique. The prepared TiO_2 nanoparticles show tetragonal rutile TiO_2 structure and average size is 10 nm. The plausible mechanism of TiO_2 nanoparticles from M. citrifolia leaves extract is given in the accompanying Figure 3. Additionally, they infer that the spherical shape of TiO_2 nanoparticles created a absorption peak at 350 nm, demonstrating the stability of TiO2 nanoparticles.

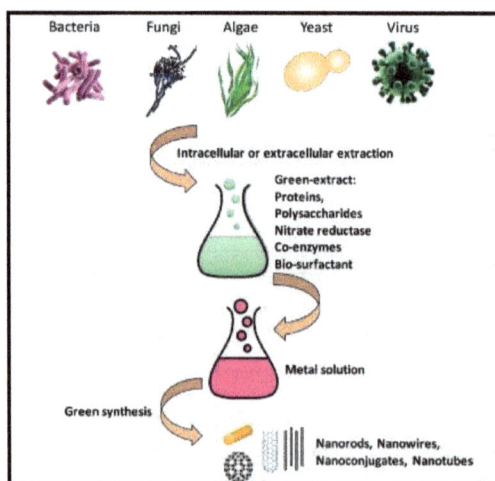

Figure 3 shows graphical representation of metal nanoparticles synthesis from microbes (Copyright from [39])

3.   Bio-mediated synthesis of nanoparticles for diverse applications

Application of the bio mediated synthesis in nanoscience and nanotechnology that maximize societal benefit and minimizing impact on the environmental remediation [40]. The bio mediated metal and metal oxides semiconductors are efficient in potential areas such as photovoltaic, thermoelectric, photocatalysis, biosensors, cosmetics, drug delivery, food preservation [41], waste water purification, and pharmaceutical etc. The bio-mediated synthesized nano materials may be useful in coating of materials such as photoanode and counter electrode in DSSC, and electrodes in energy storage and conversion devices etc. The detailed applications of bio-mediated metal oxide nanoparticle are described in Figure 4. Ag nanoparticles are impregnated with microorganisms, these are widely used in medical applications such as cardiovascular and bone implants (regenerations).

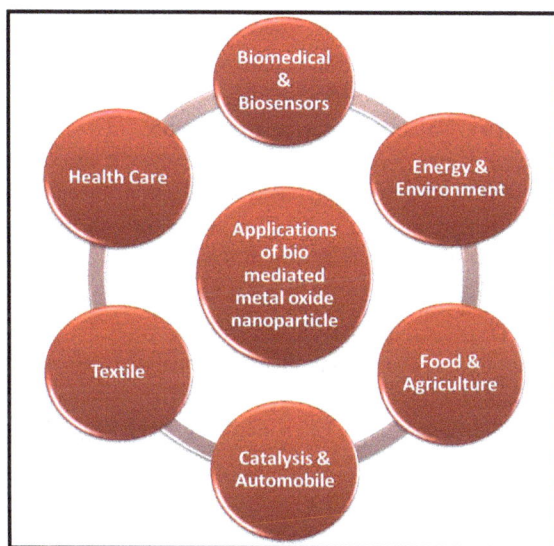

Figure 4. An overview applications of Bio-Mediated metal oxide nanoparticles

4.   Overview of dye sensitized solar cells

Dye-sensitized solar cell (DSSC) is a third-generation photoconversion device that has concentrated by researchers a tremendous attention for the past few decades, and it's an alternative of silicon and thin film solar cells. In DSSC, dye loaded photo-electrode plays a vital role in efficient light harvesting and transforming light into electrical energy at low cost. Similarly, low-cost photovoltaic panels are used for power generations windows; roof of the buildings, automobile vehicles, portable electronics and indoor light harvesting [42]. Third-generation devices are based on nanomaterials mixture of organic-inorganic hybrid materials. To the best of our knowledge the greatest power conversion efficiency (PCE) of DSSC is up to 11-13% with combination of a TiO_2 electrode, Ru based dye, and an iodine-based electrolyte [3]. The photocurrent in DSSC is produced between the junction of dye-loaded photoanode and cathod upon light irradiation. The important property of the metal-complex sensitizers is important to absorb all photons energy in the visible-NIR region, high ε, and good electrochemical stability. The metal oxide semiconductor film is functionalized with carboxylate, hydroxamate groups. The energy level of the excited state should be matched with the lower edge of conduction band and long-life time [43].

Various researchers reported that, N3 dye shows good absorptions in the visible region and permittivity (ε) of 1.35×10^4 M^{-1} cm^{-1} in alkaline medium at 500 nm. Similarly, N719 has ($\varepsilon = 1.42 \times 10^4$ M^{-1} cm^{-1}) N749 ($\varepsilon = 7.7 \times 10^3$ M^{-1} cm^{-1}). These three dyes have SCN$^-$ molecular geometry, broad absorption in visible-NIR region, and in brown-black colour and also recorded good conversion efficiency (η) = 10 and 11.1%, for N3 & N749 and similar V_{OC} (~720 mV). To achieve device performance, new approaches to turn the different parameters or strong research efforts on the dyes such as dye photoresponse, redox potential of sensitizer, permittivity, efficient electron injection, reduction of dye aggregation etc. Search alternative metal free and green dyes for the large scale development of DSSC is as important as search for different dyes for sensitizer and this issue is discussed in the discussed in the following section.

The advancement of DSSC is more flexible, offers good optical properties including light weight, optical transparency, functioning at low intensity or indoor light, and easy for installation in buildings as solar or decorative windows. The best organic DSSC installation was made in SwissTech Convention Center, Ecole Polytechnique Fédérale de Lausanne (EPFL), Lausanne, Switzerland.

However, the efficiency of DSSC has not reached the predicted theoretical values due to the loss and recombination of charge carriers during the transport and therefore researchers are working towards the following key factors.

> ➢ Good electron-transport characteristic materials for efficient charge collection and composite structures for stable photoanodes

> ➢ Excellent light absorbing property

> ➢ Design and development of excellent dye for absorbing wide region of light

> ➢ Combinations of two or more sensitizers for an effective route to extend light absorption

➢ Improving porous nature of electrode materials for sufficient dye loading

➢ Acceptable porosity of the film for feasible diffusion of the electrolyte

To overcome the above said key factors, one dimensional (1D) nanostructure enhances the dye loading efficiency in improving light-harvesting capacity, and in retain higher electron-transport.

A regular structure of DSSC contains a photoanode, a counter electrode, and an electrolyte. The photoanode as the vital piece of cell affords electron transport among sensitizer and external circuit. The light harvesting efficiency (LHE) is given by $LHE(\lambda)=1\text{-}10^{-\alpha d}$, d is the thickness of the film (10 μm thick film is the optimum thickness for an efficient collection of electrons), α is the optical absorption cross-section of the sensitizer ($\alpha=\sigma c$) α is concentration of dye loading on photoanode film and c is light velocity. The absence of porous in films is not supported to sufficient dye loading because the electron diffusion path $L_n=\sqrt{D_e\tau}$ is limited, where τ is electrons life time and D_e is the charge carrier's diffusion coefficient, respectively. On account of these issues, researchers are working on shape-persistent templates, an advantage of this template is maintained throughout the synthesis route as it will not affect the reaction.

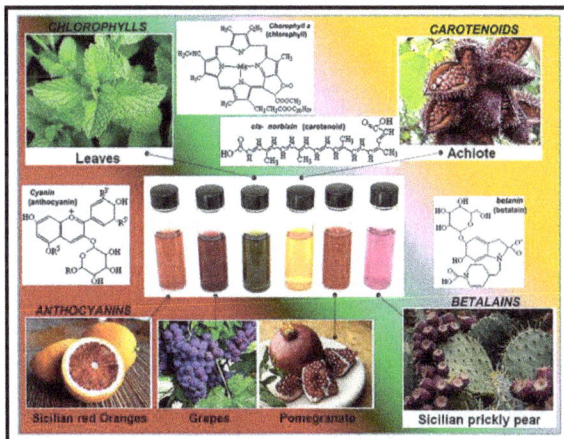

Figure 5. shows overview of dyes extracted from chlorophylls, anthocyanins, betalains and carotenoids exhibiting color covering the whole visible spectrum [44].

Bioinspired Nanomaterials for Energy and Environmental Applications   Materials Research Forum LLC
Materials Research Foundations **121** (2022) 175-210   https://doi.org/10.21741/9781644901830-6

Organic dye extracted from flowers, fruit, leaves, algae, can be used as good sensitizers in DSSCs. Most of the researchers are working on different dyes with different metal oxides. The organic derivative dyes and natural dyes are excellent candidates for environmentally friendly solar cells due to the abundant, low cost and nontoxic. In 2014 Giuseppe Calogero et al [44] described about various dyes extracts from chlorophylls (chloros = green and phyllon = leaf), anthocyanins (anthos = flower and kianos = blue), betalains (Beta vulgaris) and carotenoids exhibiting entire visible region as shown in Figure 5.

Figure 6 shows the overview, working principles of DSSC and photochemical process upon optical excitation. Metal oxides are a generally excellent optical material that ought to transmit the whole visible-near IR into the solar cells. In the initial process light strikes the dye molecules present in the TiO_2 composites. The photosensitizer absorbs the electromagnetic spectrum and then pumps excited carriers photoanode in the first process (Eq. (1)), i.e., photoexcitation of the dye molecules takes place resulting in the excitation of an electron into the conduction band [1] in typical time constant k, around 10^{-12} s, under one sun light.

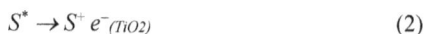

$$S + h\nu \rightarrow S^* \tag{1}$$

$$S^* \rightarrow S^+ \, e^-_{(TiO2)} \tag{2}$$

During the second process (Eq. (2)), electron is injected into the conduction band of the metal oxide, electron injection pathway is improved, when the dye is emphatically moored onto the semiconductor surface.

Thus, injected electron travels through the semiconductor and finally reaches conductive substrate in the third process, and also accepts an electron from the charge mediator and repeats the cycle again and again. The increasing charge carried density in the photo anode (TiO_2) results enhance the electrochemical potential difference (ΔV) between the TiO_2 and electrolyte. In the fourth stage, the dye-molecules are regenerated from their ground state by capturing electrons from the electrolyte. During the fifth stage, the I_3^- ions are formed by oxidation of I^- diffusing through the electrolyte to the sandwiched cathode, where the regenerative cycle is completed by electron transfer to reduce I_3^- to I^-.

In the sixth stage, the injection process happens is accepted in the femtosecond time scale for a Ru-complex sensitizer connected to an oxide surface (recover the oxidized sensitizer. This is given by the excited state lifetime of the dye, which for ordinary Ru-complexes used in DSSCs are 20–60 ns. In seventh stage, the kinetics of the back-electron transfer reaction were discussed from the semiconductor conduction band to the oxidized sensitizer in Eq. (3)

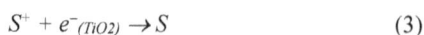

$$S^+ + e^-_{(TiO2)} \rightarrow S \tag{3}$$

$$I_3^- + 2e^-_{(TiO2)} \rightarrow 3I^-_{(anode)} \qquad (4)$$

In the final stage (8), the recombination of electrons in TiO_2 with acceptors in the electrolyte, Eq. (4) is usually referred to as the electron lifetime (τ_n, 1-20 ms). Thus, the optimization of DSSCs requires looking at the energetic levels of dye (dye regeneration processes), conduction band, electrolyte and counter electrode respectively. Finally, the counter electrode will play a major role for charge mediator otherwise the device efficiency is limited. Also, the redox potential is a potential operation of device, where the LUMO of dye molecules must have energy higher than oxide conduction band and the HOMO ought to be lower regarding redox potential. Electrical properties of the sunlight-based cells have an inherent electric field which gives the required voltage to drive the flow through an outside burden. Figure 6 gives detailed explanation on the fundamental processes and principles of DSSCs. In-depth knowledge and understanding of the electron-transfer kinetics is key role in identifying materials for DSSCs.

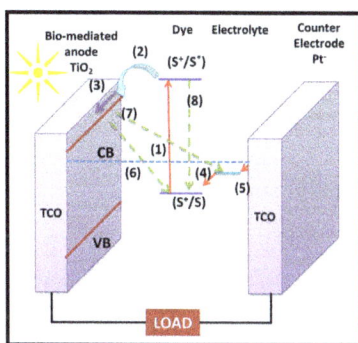

Figure 6. Illustrates the fundamental processes and working principles of DSSCs

In the device performance the short-circuit current density (J_{SC}) is generated per unit area (mA cm^{-2}) of the cell under sun light. The J_{SC} is dependence on the capability of the sensitizer to absorb light in the solar spectrum and inject electrons in the semiconductor. Moreover, the J_{SC} value mainly depends on the reduction rate of the oxidized dye molecules. If it is lower, the dark current will be increased, and the J_{SC} will be reduced. The open circuit voltage (V_{OC}) is the electrical potential difference at open circuit between the two terminals of the cell under light conditions. The greatest V_{OC} estimation of a DSSC (0.8–0.9 V) corresponds to the difference between the quasi-Fermi level (EF, - 0.5 to - 0.4 eV versus typical hydrogen electrode - NHE, for TiO_2) of the semiconductor and the redox potential of the charge mediator (- 0.4 V versus NHE, for I_3^-/I^- redox couple) [45]. Indeed, for regenerative photo- electrochemical systems.

$$V_{oc} = \left(\frac{K_B T}{e} \right) \ln \left(\frac{I_{inj}}{n_{cb} k_{red} \left[I_3^- \right]} \right) \tag{5}$$

where k_B is the Boltzmann's constant, T is the absolute temperature, e is the charge of electron, I_{inj} is the charge flux resulting from sensitized injection, k_{red} is the reduction rate constant of $I_3^- n_{cb}$ is the concentration of electrons at the semiconductor surface and I_3^- is the oxidized redox mediator in the (I^-/I_3^-) solution.

Fill Factor (FF) is defined by following formula

$$FF = \frac{P_{max}}{J_{SC}.V_{OC}} \tag{6}$$

where P_{max} is the maximum power output of the cell per unit area, ($J_{max} \times V_{max}$). In current–voltage characteristics, the FF is evaluated under the area of current voltage curve fills-in the maximum rectangle by $J_{SC} \times V_{OC}$. The shape of this curve is a tool to evaluate the DSSC performance and it is shown in Figure 7 (i.e., V_{OC}, J_{SC} and η). The energy conversion efficiency (η) of the solar cell is defined as the ratio between the output electrical power (P_{max}) and the energy of the incident light (P_{light}) and is given by the following relation:

$$\eta = \frac{P_{max}}{P_{Light}} = \frac{V_{OC}.J_{SC}.FF}{P_{Light}} \times 100 \tag{7}$$

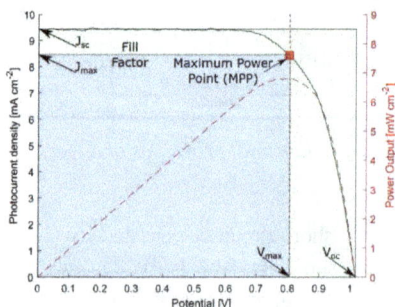

Figure 7. Current voltage characteristics of solar cell to evaluate the efficiency of the device (Copyright from [2])

External quantum efficiency (EQE) of the solar cells can be defined as the quantum efficiency of the cell at a given excitation wavelength and corresponds to the photocurrent density generated in the external circuit per incident monochromatic photons flux that strikes the cell.

$$EQE(\lambda) = \frac{N_e}{N_p} = 1240 \times \frac{J_{SC}(\lambda)}{P_{Light}\lambda} \qquad (8)$$

$$EQE(\lambda) = LHE\phi E1\eta EC \qquad (9)$$

where, LHE is the light harvesting efficiency (The detailed LHE is explained in the first part), $\phi E1$ is electron injection quantum efficiency, and ηEC is the efficiency of the device. Limitation towards increase the efficiency of the device is the stability (stability of the electrolyte is more than 80°C, because during the sunny days temperature goes to the maximum and in normal days it will reach 60-70° reasonable temperatures at AM1.5 also stable for 1000 hr). Overview of DSSC module is shown in Figure 8. Since 1991, huge financial and scholarly ventures have been made to form DSSCs into deployable advances, making an abundance of information about nano-interfaces and devices through an expanding number of exploration reports. Since DSSCs is a simple manufacturing and cost-effective photovoltaic device, it is still evaluating the various parameters and electrode materials for further improving performance. In this regards the content of this book chapter mainly has focused on role of the bio-mediated synthesis nanoparticles on DSSCs performance.

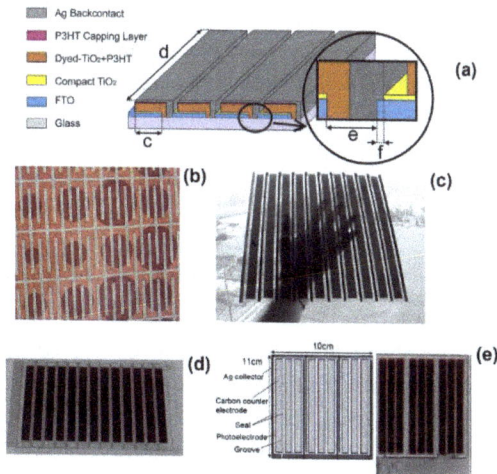

Figure 8. Schematic and photographic view of DSSC module (Copyright from [42])

Bioinspired Nanomaterials for Energy and Environmental Applications   Materials Research Forum LLC
Materials Research Foundations **121** (2022) 175-210   https://doi.org/10.21741/9781644901830-6

5.   Bio-synthesized nanoparticles for dye-sensitized solar cell application

Various synthetic methods have been used for preparation of the nanomaterials for tuning morphology, structure and uniformity and size. Even though the methods have shown better results but still a development or modifying the methods is highly important which could be exploited at the modern and business level to have better grown, durable, long lasting, cleaner, more secure and more smarter products for example, home appliances, communication innovation technology, medicines, transportation, agriculture and enterprises. So, focus on design and preparation of the nanomaterials using environmentally benign approaches is very important.

The synthesis of metal, metal oxide, metal sulphide, core-shell nanoparticles and nanocomposites using biological elements has received great attention due to their interesting size depending properties. The bio-mediated synthesized nanoparticles were applied various applications particularly, energy storage and energy conversions. The bio-mediated synthesized metal and metal oxides nanoparticles were successfully used in dye-sensitized solar cells in various parts of the devices.

Due to the wide-band gap and efficient electron transport material, the TiO_2 has been used extensively as photoanode for DSSCs fabrication [1]. The synthesis of TiO_2 with bio-mediated method for solar cells and opens the interest in energy devices as a way to develop greener photovoltaic technologies. Several authors have reported bio-mediated synthesized ZnO nanoparticles for various applications including, photovoltaics, biosensors for cholesterol, enzyme biochemistry, and other biosensing applications [46]. Apart from the TiO_2 and ZnO electron transport materials, several other metal oxides such as CuO, CoO and SnO_2 were prepared by the green method. Moreover, Ag, Au, CdS and ZnS nanoparticles were prepared and successfully applied as sensitizer in the DSSCs. The following sections briefly discuss types of the nanoparticles synthesized through the bio-mediated route for the DSSCs applications.

6.   Bio-mediated approach for the synthesis of metal nanoparticles for DSSCs application

Metal nanoparticles can play to the ground-breaking light absorbing in sun light, both by local field upgraded through the limited surface plasmon resonation and by light disseminating prompting delayed optical-way lengths [47]. Plasmonic NPs show two trademark properties when they associate with visible light. Initially, they show an extraordinary absorption feature at specific wavelength of light which are resonant with their electron oscillation.

Bioinspired Nanomaterials for Energy and Environmental Applications   Materials Research Forum LLC
Materials Research Foundations **121** (2022) 175-210   https://doi.org/10.21741/9781644901830-6

In the preparation of metallic nanoparticles utilizing plant extracts, it just mixes with the precursor metal salt at room temperature and the reaction is finished in a few minutes [48]. Metal nanoparticles of silver, gold, platinum and many other metals have been synthesized in this way for DSSCs application to use in the different parts of the device. The parameters such as concentration of the extract, metal salt, the pH of the solution, temperature of the reaction solution and production medium are mainly affecting the rate of the nanoparticles production, their quantity and quality. Adding of noble metals such as silver or gold nanoparticles on TiO_2 surface is one of the best ways of modifying the photoanode of DSSCs. It greatly influences the optical and electronic properties of TiO_2.

Recently plasmon resonance effect has been brought into the DSSC using metal particles, for example, silver and gold [49]. It is one of the techniques developed to improve the productivity of DSSC. The localized surface plasmon resonance (LSPR) effects of the nobel metals increase the light harvesting efficacy. The LSPR refers to the resonance between the electromagnetic field and free-electron oscillation. It intensifies the electromagnetic field close to the metal nanoparticles, coming about in plasmon upgraded light absorption by dye sensitizers in DSSC. As of now, silver nanoparticle-enabled plasmonic effect has been proposed and shown as a promising method to manage achieve light catching in solar cells. In this method, the relative scattering yield of silver nanoparticles is higher than that of other noble metals in the visible region. [44]. This material has numerous advantages for sensitized photochemistry and photoelectrochemistry in light of the fact that it is an ease, non-poisonous and biocompatible material, and as such it is used in medical services items just as domestic applications, for example, paint pigmentation.

This topic is mainly focusing on the DSSCs fabricated using noble metal nanoparticles as these metals nanoparticles were synthesized through green route. Figure 9 shows bio-mediated synthesis of metal nanoparticles from plant leaf and decoration of the metal nanoparticles with TiO_2 and then fabrication DSSC using the TiO_2-metal nanoparticles.

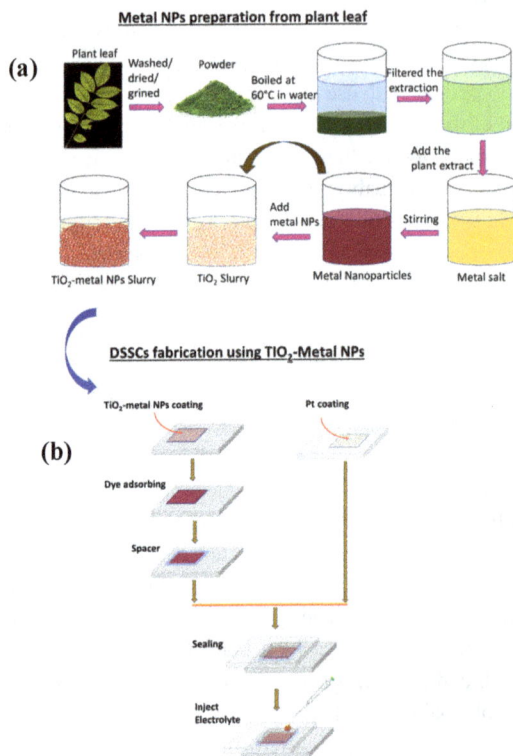

Figure 9. (a) Bio-mediated metal NPs decorated TiO$_2$ photoanode and (b) DSSC fabrication

6.1   Ag Nanoparticles

Modification of the photoanode (TiO$_2$) with Ag is an effective way for improving DSSC's efficiency. This metal decoration is greatly influencing in the electronic properties of TiO$_2$. Figure 10 shows graphic representation of the Ag decorated TiO$_2$ anode preparation. Numerous results were reported on the green synthesis of Ag nanoparticles and a few researcher groups fabricated DSSCs using the Ag nanoparticles. Since the green synthesis method is simple and low cost, researchers are giving more attention to the method. Moreover, the Ag nanoparticles are low toxic. For preparation of the Ag mixed with TiO$_2$, ethanol was used as dispersing medium.

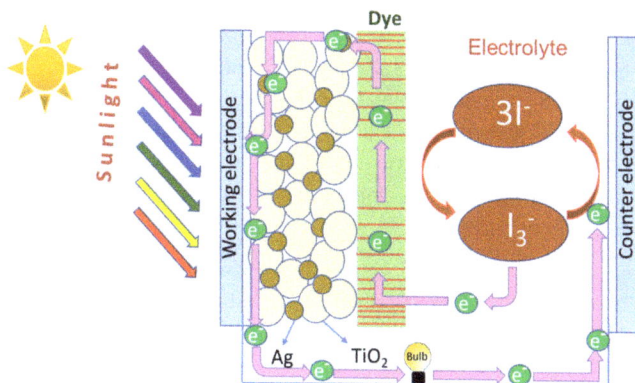

Figure 10. Schematic illustration of Ag decorated TiO₂ photoanode with dye sensitized solar cell

Recently, Saravanan et al. [50] described that power conversion efficiency of DSSCs has been amazingly increased by incorporation of bio mediate synthesized silver nanoparticles into TiO₂ photoanodes [50]. Due to the plasmonic effect of the altered cathode, the efficiency of the DSSC was discovered to be improved from 2.83% to 3.62% with increase around 28% after adding of the 2 wt.% of the silver nanoparticles. They found an increasing in Voc (up to 12.1%) and Jsc (up to 10.7%).

Danial et al. [51] prepared silver nanoparticles using Henna (Lawsonia inermis) leaf extract at room temperature. Size of the silver nanoparticles was controlled by leaf extract concentration, pH, reaction temperature on the development of nanoparticles were evaluated by the varying in absorbance and wavelength of surface plasmon resonance of the nanoparticles [51]. The particles of 50 nm had been confirmed by AFM. The silver NPs present with Henna leaf extricate which contains the dye molecules of lawsonine were utilized as a photosensitizer in a photo improved DSSC. High output of photocurrent of a photo enhanced – DSSC (44 mV) was obtained compared to DSSC (32 mV). The enhancement of the photovoltaic parameters of DSSCs is due to the LSPR effect of silver nanoparticles.

6.2   Au nanoparticles

Numerous researchers showed that the size of the metal nanoparticles is a key factor for deciding the plasmonic phenomena, and particularly, light scattering happens more dominantly than light absorption as the size of the metal nanoparticles increments. [52]. Nevertheless, Chang woo Nahm et al [53] prepared gold nanoparticles with 100 nm size. The Au nanoparticles were incorporated into TiO_2 nanoparticles for DSSC. They optimized Au/TiO_2 mass proportion of 0.05, and the PCE of the DSSC improved to 3.3% from an estimation of 2.7% without Au, and they claimed that this enhancement was mostly ascribed to the photocurrent density. In addition, they concluded that enhancement of the efficiency for the Au nanoparticles embedded TiO_2 film is emphatically absorbed light due to surface-plasmon resonance.

Solaiyammal et al. [54] prepared gold nanoparticles through a bio mediated synthesis method in aqueous extract from the phyllanthus embilica (commonly known as emblic or amla or thoppunelli) leaves without using any surfactant and stabilizing agent [54]. They used pristine TiO_2 and Au incorporated TiO_2 nanocomposite as photoanode for the fabrication of the DSSCs. The $Au:TiO_2$ nanocomposites prepared by mixing the gold nanoparticles with TiO_2 slurry. The $Au:TiO_2$ nanocomposite showed about 65% improvement of the DSSC efficiency compared to the pure TiO_2. The physical parameters such as charge transport resistance and life time of the charge carries was evaluated by electrochemical impedance spectroscopy. The incorporation of green synthesized gold nanoparticles with TiO_2 ensuring significant improvement of the current density followed by the device performance. Therefore, the research group concluded that the overall PCE improvement from 5.2% to 8.6% of Au incorporated TiO_2 based nanocomposite is correlated with pristine TiO_2 in DSSC [54]. Moreover, they claimed that surface plasmon resonance effect of gold and the development of Schottky barrier at the interface of Au and TiO_2 were the reasons behind the improvement of the efficiency of $Au:TiO_2$ nanocomposite based DSSC. Accordingly the lesser charge transfer resistance (Rct) in $Au:TiO_2$ photoanode is corresponded to that of pristine TiO_2 photoanode prompting an enhancement in the performance of $Au:TiO_2$ photoanode DSSC. In addition, electron life time was found from the Bode stage plot and the high electron lifetime (4.35 ms) was found for $Au:TiO_2$ photoanode based DSSC related to that of 2.84 ms for the pristine TiO_2 based DSSC.

Apart from the Ag and Au nanoparticles, some other metal nanoparticles were synthesized by bio-mediated process for DSSCs applications. For example, Asghar et al [55] were prepared various metal nanoparticle such as Zn, Zr and Cu with size of 4-15 nm. The prepared metal nanoparticles were mixed with TiO_2 and used the metal nanoparticles decorated photoanode. The PCE for 40 mol% of Zn, 5 mol % of Zr and 5 mol % of Cu with titania were found to be 4.8, 4.95 and 4.9 %, respectively.

7. Bio-mediated synthesised metaloxide nanoparticles for DSSCs application

Metal oxide nanoparticles have high surface to volume ratio which is liable for their captivating properties, for example, antimicrobial, synergist action, electronic, magnetic, photovoltaic and optoelectronics. Generally, metal oxide nanoparticles were incorporated by different physical and chemical methods including-sputtering, reduction, sol-gel, hydrothermal and electrochemical techniques. However, these techniques are expensive, high pressure, toxic, high energy necessity and potentially hazardous. Plant extract for the synthesis of the metal oxide nanoparticles is considered to be the simple and rapid process. Moreover, the bio-mediated synthesis is reliable and eco-friendly method. Various metal oxide nanoparticles such as TiO_2, ZnO, SnO_2, CuO, Co_3O_4, ZrO, Fe_3O_4 and WO_3 were prepared using the bio-mediated method for dye-sensitized solar cells applications. In this context different metal oxide nanoparticles prepared by bio-mediated synthesis for DSSC application are discussed below.

TiO_2 is the most known photo anode materials for DSSCs. It plays a vital role DSSC, due to its small particle size, high surface area, more active anatase phase, high band gap energy and high electron mobilities. [56]. TiO_2 is an n-type semiconductor with an energy band gap of 3.2 eV. The biological synthetic method is affiliated with the use of extracts from living organisms and has been reported to be the most ecological friendly approach for the preparation of TiO_2 NPs because it does not involve the use of toxic chemicals, neither does it release toxic chemical products to the environment during the synthetic process. The following some plants extracts such as Eclipta Prostrate Extract, Nyctanthes Arbor-Tristis, Vigna Radiata Extract, Psidium Guajava Extract, Fusarium Oxporum Extract, Arnicae Anthodium Extract, Curcuma Longa Extract and Bacillus Mycoides were mostly used for the preparation of TiO_2 nanoparticles. Accordingly, Ekar et al [57] were prepared. Nanorods of TiO_2 have biogenically from the extract of Phellinus linteus mushroom. They confirmed the mixed structure of anatase and rutile TiO_2 phases. The DSSC is fabricated using the bio mediated synthesized TiO_2 photoanode in the presence of N719 dye prepared in 0.5 M lithium iodide and 0.05 M iodine with acetonitrile as redox electrolyte. The DSSC shows a short out current density (Jsc) of 8.18 mA/cm^2, an open-circuit voltage (Voc) of 0.69 V, a fill factor (FF) of 67 %, and a PCE (η%) of 3.80 %.

Recently, Maurya et al. [58] prepared mesoporous anatase TiO_2 nanoparticles using Fresh seeds of *Bixa orellana* (*B. orellana*). The preparation method is clearly demonstrated in Figure 11. They prepared mesoporous TiO_2 nanoparticles using titanium (IV) butoxide and *B. orellana* extract. For comparison purpose the TiO_2 NPs were prepared by a conventional method as shown in Figure 11.

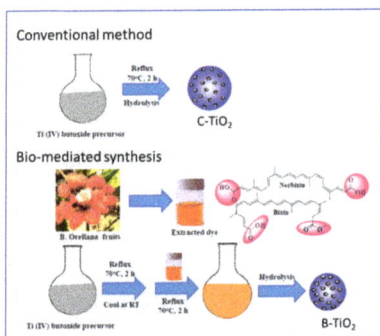

Figure 11. Schematic illustrations for synthesis of TiO₂ nanoparticles using conventional sol-gel method (C-TiO₂) and with the use of B. orellana extract (B-TiO₂) (Copyright from [58]).

At first, 10 ml of titanium butoxide was included 80 ml isopropanol and the mixed solution was then refluxed for 4 h at 70 °C to get clear solution. It was cooled to room temperature followed by the expansion of 5 ml glacial acidic acid to get solution A. The solution A refluxed for 2 h at 70 °C and afterward cooled to room temperature followed by addition of 5 ml of natural dye extract. This was followed by refluxing for 2 h at 70 °C, afterward cooled down to room temperature. Then, solution A was included drop wise into arrangement B (15 ml of deionized water) under vigorous stirring. The mixed solution was stirred constantly for 2 h and aged for 24 h at room temperature. Then, the resultant precipitates were recovered by centrifugation and were washed thoroughly with ethanol and deionized water followed by drying in air oven at 60 °C for 24 h. At last, the powder was calcined at 450 °C for 1 h. The obtained mesoporous anatase TiO₂ nanoparticles were utilized as photoanode materials for DSSC fabrication. The morphology (SEM and TEM pictures) of the TiO₂ nanoparticles prepared by both traditional and bio-intervened techniques are displayed in Fig. 12(a-d). The amount of dye stacking by TiO₂ was gotten using a desorption procedure which revealed a much higher dye stacking for the plant seed developed nanoparticles of (G-TNP). TNP and G-TNP showed photovoltaic transformation efficiencies 1.03% and 2.97% individually. This investigation hence exhibits that the green synthesis TiO₂ nanoparticles can serve as promising photoanode materials for DSSC applications.

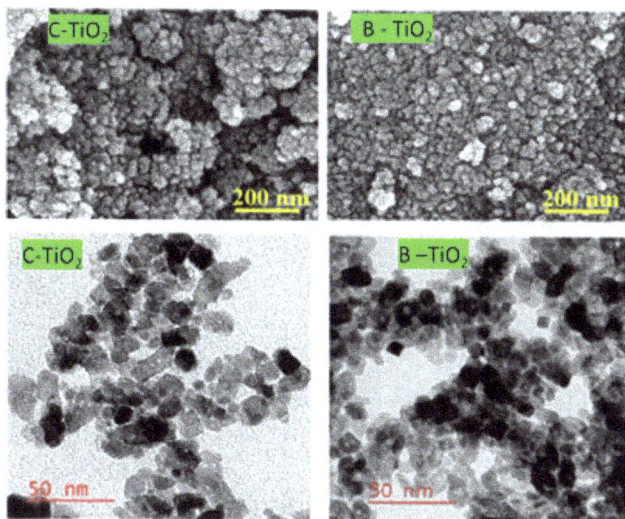

Figure 12. HR-SEM and HR-TEM images of TiO₂ nanoparticles using conventional sol-gel method (C-TiO₂) and with the use of B. orellana extract (B-TiO₂). Apart from the TiO₂ NPs, ZnO, CuO, Co₃O₄, SnO₂ and ZrO₂ nanoparticles were also prepared by bio-mediated technique (Copyright from [58]).

Rajeswari et al [59] reported a simple method for synthesis of hexagonal wurtzite ZnO nanoparticles through green method utilizing Carica papaya leaf extract. The obtained result showed that the synthesized ZnO nanoparticle is a pure hexagonal wurtzite structure with spherical shape with size of ~50 nm. It was found that the prepared ZnO nanoparticles used in DSSCs and it demonstrated the PCE of 1.6% with a current thickness of 8.1 mA cm^{-2} in DSSCs. Sharma et al. [60] have reported bio mediate synthesized copper oxide (CuO) NPs as counter electrode in DSSCs. The CuO NPs were prepared by the leaves extract of Calotropis gigantean plant in aqueous medium through green synthesis method. The synthesized CuO NPs had a well crystalline nature with monoclinic structure of CuO. For DSSC application, a slim film of synthesized CuO NPs was prepared by the paste of CuO NPs with ethanolic nafion solution and spread onto FTO glass utilizing glass rod. Moderately, high solar energy to electrical energy efficiency of ~3.4% alongside high short current density (J_{SC}) of ~8.13 mA/cm², open circuit voltage (V_{OC}) of ~ 0.676 V and fill factor (FF) of 0.62 was recorded in the DSSC fabricated with prepared CuO NPs based counter cathode. The authors were concluded that the biosynthesized CuO NPs counter cathode may deliver countless dynamic destinations for the high reduction of I_3^- to I^- ions and improve the conductive connection for the charge transfer, bringing about the high J_{SC}, FF and conversion efficiency.

Bio-mediated synthesis of cobalt oxide (Co_3O_4) NPs using leaves extract of plant calotropis gigantea were reported by Sharma et al [61]. The image of plant leaf, extract, paste formation and annealed Co_3O_4 NPs are presented in Figure 13(a-e). In addition, the FE-SEM image, EDX and TEM images of the green synthesized NPs are showed in Figure 14(a-d). The obtained device performance ($Jsc = 1.79$ Ma/cm2; $Voc = 0.646$V, FF = 0.57) is quite low as compared to Pt based DSSCs, but the green synthesized Co_3O_4 NPs as new counter electrode materials show good electrocatalytic activity for the reduction of I_3^- to I^- ions.

Figure 13. Photograph of (a) C. gigantea, (b) leaves extract, (c) reaction mixture, (d) as synthesized past and (e) annealed Co_3O_4 NPs (Copyright from [61])

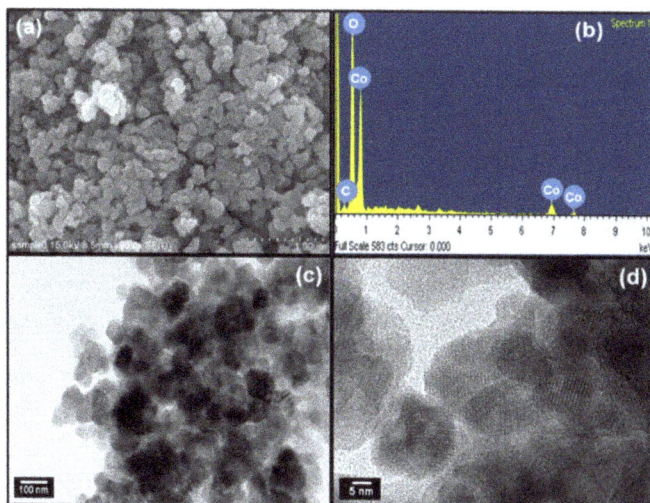

Figure 14. (a) FE-SEM image, (b) EDX spectrum, (c) low magnification HR-TEM image and (d) high resolution image of Co_3O_4 NPs (Copyright from [61])

Wu et al. [62] synthesized successfully the WO_2 nanorod which consisted of beef and was used in DSSCs as a counter electrode and obtained a high PCE of 7.25%. Similarly, Liu et al. [63] prepared $W_{18}O_{49}$, W_2N, WC, and WO_3 based counter electrode materials in DSSCs systems and demonstrated superior PCE of 6.69%, 5.97%, 5.20%, and 4.69% respectively in the I_3^-/I^- redox shuttle, [63]. Recently, Li et al. [64] were reported high power conversion efficiency using WO_3 based counter in DSSC. DSSCs with zinc tungstate-carbon CE catalyst rendered an overwhelming PCE of 7.61%, comparable to Pt-sputtered CE (7.04%). Manganese tungstate-carbon and coper tungstate-carbon electrodes in DSSCs yielded favorable PCE of 7.33% and 6.52%, respectively. The synthesis of biomas-derived carbon nanoparticles is clearly demonstrated in Figure 15.

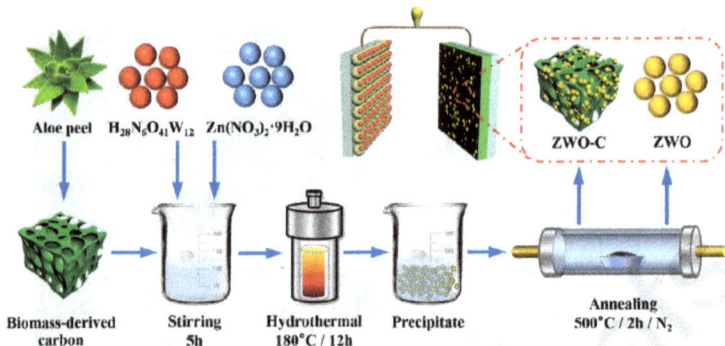

Figure 15. Demonstrated biomass-derived carbon and followed by the carbon-tungsten oxide nanocomposite for counter electrode in DSSC (Copyright from [64]).

8.   Bio-synthesized of Core-shell/QDs on DSSCs Application

Bio-mediated synthesis is cost-effective, simple and environmentally friendly, hence widely applied for the synthesis of metal and metal oxide nanoparticles [65]. However, morphologically controlled synthesis of core-shell in biosynthesis method is very rare and mainly spherical NPs or a mixture of shapes with a wide size distribution are formed. Chemically processed double metal oxide-based core-shell nanoparticle (ZnO/TiO_2, ZnO/Nb_2O_5 etc.,) rarely used for DSSC applications [66]. However, metal covered metal oxides such as SiO_2/Au, TiO_2/Ag, TiO_2/Pt, TiO_2/Au and metal sulphide-based core-shell quantum dots (QDs) such as CdS/CdSe, CdTe/CdS, ZnSe/CdS and PbS/CdS QDs were extensively used for the DSSCs applications [67]. But the synthesis of core-shell nanoparticles and quantum dots are limited by bio mediated technique. Among all nanoparticles classified by their size, QDs arise as the key semiconductor material in the emerging nanotechnology.

Gao and his co-workers [68] prepared CdSe quantum dots by green synthesis for quantum-dots-sensitized solar cells (QDSCs) application. 3-Mercaptopropionic acid capped water-dispersible QDs were used to cover the surface of TiO_2. The resulting green synthesized CdSe QDSCs with Cu_2S as the cathode show a photovoltaic performance with conversion efficiency. The conversion efficiency of the green synthesized CdSe QDSCs reached 3.39% (Voc = 0.549 V, Jsc = 11.86 mA/cm^2, FF = 0.52). The authors suggested that the efficiency enhancement is due to suitable panchromatic QD sensitizers for expanding the harvest of solar light and is the key factor to improve the efficiency of the solar cell. The reported result clearly showed fabrication of high efficiency DSSCs using the green synthesized QDs.

Recently, carbon QDs have considerable advantages as they are metal-free and low cost, low toxicity, high biocompatibility and are eco-friendly, and also demonstrate a strong PL emission. A variety of applications in biosensors, bioimaging, elemental sensors, medicine as well as in catalysis have been demonstrated [69]. Guo et al [70] prepared CQDs from several carbon sources for DSSCs application. The carbon QDs (CQDs) were synthesized from three different carbon sources, bee pollen, citric acid, and glucose, through a hydrothermal route. The fabricated CQDs sensitized solar cells achieved the power conversion efficiency of 0.11%. CQDs derived from such a simple and green method may provide us with a new perspective towards their application in a variety of fields. Undeniably, the efficiency accomplished using CQDs now is not satisfying but it can be potentially improved by further refining the particle size and functional groups on the surface.

Summary and conclusion

Dye sensitized solar cells (DSSC) are moderately new class of cost-effective solar cells that belong to the category of thin film solar cells. They are promising technology compared with silicon-based solar cells as they are produced low-cost materials, and don't require advanced methods and machine to be manufactured metal oxide nanoparticles, for example, tin oxide, iron oxide, zinc oxide and titanium are interesting and play a critical part in different device applications. Among the different metal oxide nanoparticles, titanium dioxide nanoparticles have wide applications for DSSCs, air and water purification, because of its potential oxidation quality, high photostability and non-poisonousness. Till now, titanium dioxide (TiO_2) is the foundation semiconductors for DSSC nanostructured anodes for DSSCs.

There is a need of additional examination on the biosynthesis mechanism for TiO_2 nanoparticles because that they have possible applications in coming years. Bio mediated synthesis of TiO_2 from natural source is the most ideal approach to use so as to get enough amount of TiO_2 for DSSCs as well as other various applications. The utilization of biological sources, for example, plant, microbes, yeast and fungi for the preparation of TiO_2 nanoparticles is ease, low-cost and simple process. In this regard this book chapter has elaborately discussed the synthesis of TiO_2 nanoparticles using various plant extracts and the performance of DSSCs has been found to be similar to that of the TiO_2 synthesized by other methods.

Next, the ZnO nanoparticles were effectively synthesized by a simple and eco-friendly bio-mediated green method utilizing different plant leaves. The ZnO nanoparticle-based DSSC performance showed also similar to that of other methods. The high performance of the DSSC is because of a significant increment in the absorption of dye molecules on the surface of ZnO nanoparticles. In this way, the utilization of green synthesized ZnO nanoparticles in fabrication on DSSC is a simple and promising technique for the wellness of our future.

Fascinatingly, the solar cell application revealed that DSSC fabricated with green synthesized Co_3O_4 NPs based counter electrode accomplished sensible conversion of solar energy to electrical energy efficiency. In this way, the green synthesized Co_3O_4 NPs have demonstrated to be promising catalytic materials for electrode application in DSSC. The following ease and notable SnO_2 NPs were synthesized by green technique and utilized as electrode in DSSCs.

Moreover, the green chemistry approach demonstrated that copper oxide nanoparticles also can be synthesized by bio-mediated technique, and moreover it showed that it is simple, cost effective, highly stable and reproducible method.

The green technique for prepare CuO nanoparticles could likewise be stretched out to manufacture other, industrially significant metal oxides. Synthesized CuO NPs used as electrocatalytic materials for the preparation of counter cathode in DSSC and sensor applications.

The green technique is demonstrated that it is likewise appropriate for preparation of carbon and carbon derived nanoparticles. The reported outcomes demonstrated that biomass-derived carbon-upgraded tungsten oxide compound as counter electrode catalyst are promising options to Pt based counter electrode in DSSCs.

Green synthesized Ag nanoparticles and their assemblies found application in different fields such as sensors, antimicrobial, peptide probes wound healing agent, and in the areas of optics, magnetism, electrical engineering and catalysis. Particularly, the green synthesized Ag nanoparticles were used with TiO_2 as it is known show the plasmonic effect, resulted in improved power conversion efficiency.

Gold nanoparticles (AuNPs) were synthesized through green method with the aqueous extract from the leaves without using any surfactant and stabilizer. DSSC with Au: TiO_2 nanocomposite as the photoanode was successfully fabricated and its efficiency was correlated to that of the pure TiO_2 photoanode.

The exploration for environmentally friendly and reliable methods for nanomaterials preparation is an active topic. Preparation of nanomaterials with profoundly monodispersion, crystallinity, shape control, and narrow size dispersion play a vital role in quantum dots (QDs)- based applications.

Eventually, the synthesis of nanoparticles using green method have developed as a green alternative, for it is environment-friendliness, low-cost, simple and could be effectively scaled-up. Eco-friendly green synthesis of metal, metal oxides, carbon nanoparticles and quantum dots using plant extract sowed that the synthesized nanoparticles can be used in energy conversion and furthermore for energy storage application without any harmful chemical and offers a promising scope in nanomaterials preparation.

Reference

[1]   B. Oregan, M. Gratzel, A Low-Cost, High-Efficiency Solar-Cell Based on Dye-Sensitized Colloidal TiO_2 Films, Nature 353 (1991) 737–740. https://doi.org/10.1038/353737a0

[2]   I. Benesperi, H. Michaels, M. Freitag, The researcher's guide to solid-state dye-sensitized solar cells, J. Mater. Chem. C 6 (2018) 11903-11942. https://doi.org/10.1039/C8TC03542C

[3]   N. Sakai, T. Miyasaka, T.N. Murakami, Efficiency Enhancement of ZnO-Based Dye-Sensitized Solar Cells by Low-Temperature TiCl 4 Treatment and Dye Optimization, J. Phys. Chem. C 117 (2013) 10949–10956. https://doi.org/10.1021/jp401106u

[4]   J.L. Gardea-Torresdey, K.J. Tiemann, G. Gamez, K. Dokken, S. Tehuacanero, M. José-Yacamán, Gold nanoparticles obtained by bio-precipitation from gold(III) solutions, J. Nanoparticle Res. 1 (1999) 397–404. https://doi.org/10.1023/A:1010008915465

[5]   S. Stankic, S. Suman, F. Haque, J. Vidic, Pure and multi metal oxide nanoparticles: Synthesis, antibacterial and cytotoxic properties, J. Nanobiotechnology 14 (2016) 1–20. https://doi.org/10.1186/s12951-016-0225-6

[6]   N.C. Sharma, S. V. Sahi, S. Nath, J.G. Parsons, J.L. Gardea-Torresdey, P. Tarasankar, Synthesis of plant-mediated gold nanoparticles and catalytic role of biomatrix-embedded nanomaterials, Environ. Sci. Technol. 41 (2007) 5137–5142. https://doi.org/10.1021/es062929a

[7]   J.L. Gardea-Torresdey, E. Gomez, J.R. Peralta-Videa, J.G. Parsons, H. Troiani, M. Jose-Yacaman, Alfalfa sprouts: A natural source for the synthesis of silver nanoparticles, Langmuir 19 (2003) 1357–1361. https://doi.org/10.1021/la020835i

[8] J.L. Gardea-Torresdey, J.G. Parsons, E. Gomez, J. Peralta-Videa, H.E. Troiani, P. Santiago, M.J. Yacaman, Formation and Growth of Au Nanoparticles inside Live Alfalfa Plants, Nano Lett. 2 (2002) 397–401. https://doi.org/10.1021/nl015673+

[9] S. Li, Y. Shen, A. Xie, X. Yu, L. Qiu, L. Zhang, Q. Zhang, Green synthesis of silver nanoparticles using Capsicum annuum L. Extract, Green Chem. 9 (2007) 852–858. https://doi.org/10.1039/b615357g

[10] S.P. Chandran, M. Chaudhary, R. Pasricha, A. Ahmad, M. Sastry, Synthesis of Gold Nanotriangles and Silver Nanoparticles Using Aloe vera Plant Extract, Biotechnol. Prog. 22 (2006) 577–583. https://doi.org/10.1021/bp0501423

[11] G. Zhai, K.S. Walters, D.W. Peate, P.J.J. Alvarez, J.L. Schnoor, Transport of Gold Nanoparticles through Plasmodesmata and Precipitation of Gold Ions in Woody Poplar, Environ. Sci. Technol. Lett. 1 (2014) 146–151. https://doi.org/10.1021/ez400202b

[12] M. Sundrarajan, K. Bama, M. Bhavani, S. Jegatheeswaran, S. Ambika, A. Sangili, P. Nithya, R. Sumathi, Obtaining titanium dioxide nanoparticles with spherical shape and antimicrobial properties using M. citrifolia leaves extract by hydrothermal method, J. Photochem. Photobiol. B Biol. 171 (2017) 117–124. https://doi.org/10.1016/j.jphotobiol.2017.05.003

[13] D. Hariharan, K. Srinivasan, L.C. Nehru, Synthesis and Characterization of Tio2 Nanoparticles Using Cynodon Dactylon Leaf Extract for Antibacterial and Anticancer (A549 Cell Lines) Activity, J. Nanomedicine Res. 5 (2017) 00138. https://doi.org/10.15406/jnmr.2017.05.00138

[14] S.P. Goutam, G. Saxena, V. Singh, A.K. Yadav, R.N. Bharagava, K.B. Thapa, Green synthesis of TiO2 nanoparticles using leaf extract of Jatropha curcas L. for photocatalytic degradation of tannery wastewater, Chem. Eng. J. 336 (2018) 386–396. https://doi.org/10.1016/j.cej.2017.12.029

[15] T. Bhuyan, K. Mishra, M. Khanuja, R. Prasad, A. Varma, Biosynthesis of zinc oxide nanoparticles from Azadirachta indica for antibacterial and photocatalytic applications, Mater. Sci. Semicond. Process 32 (2015) 55–61. https://doi.org/10.1016/j.mssp.2014.12.053

[16] P. Sutradhar, M. Saha, Green synthesis of zinc oxide nanoparticles using tomato (Lycopersicon esculentum) extract and its photovoltaic application, J. Exp. Nanosci. 11 (2016) 314–327. https://doi.org/10.1080/17458080.2015.1059504

Bioinspired Nanomaterials for Energy and Environmental Applications
Materials Research Foundations **121** (2022) 175-210

Materials Research Forum LLC
https://doi.org/10.21741/9781644901830-6

[17] T.V. Surendra, S.M. Roopan, Photocatalytic and antibacterial properties of phytosynthesized CeO2 NPs using Moringa oleifera peel extract, J. Photochem. Photobiol. B Biol. 161 (2016) 122–128. https://doi.org/10.1016/j.jphotobiol.2016.05.019

[18] M. Yadi, E. Mostafavi, B. Saleh, S. Davaran, I. Aliyeva, R. Khalilov, M. Nikzamir, N. Nikzamir, A. Akbarzadeh, Y. Panahi, M. Milani, Current developments in green synthesis of metallic nanoparticles using plant extracts: a review, Artif. Cells, Nanomedicine Biotechnol. 46 (2018) S336–S343. https://doi.org/10.1080/21691401.2018.1492931

[19] V.K. Vidhu, D. Philip, Biogenic synthesis of SnO2 nanoparticles: Evaluation of antibacterial and antioxidant activities, Spectrochim. Acta - Part A Mol. Biomol. Spectrosc. 134 (2015) 372–379. https://doi.org/10.1016/j.saa.2014.06.131

[20] R. Sankar, P. Manikandan, V. Malarvizhi, T. Fathima, K.S. Shivashangari, V. Ravikumar, Green synthesis of colloidal copper oxide nanoparticles using Carica papaya and its application in photocatalytic dye degradation, Spectrochim. Acta - Part A Mol. Biomol. Spectrosc. 121 (2014) 746–750. https://doi.org/10.1016/j.saa.2013.12.020

[21] A.Y. Ghidan, T.M. Al-Antary, A.M. Awwad, Green synthesis of copper oxide nanoparticles using Punica granatum peels extract: Effect on green peach Aphid, Environ. Nanotechnology, Monit. Manag. 6 (2016) 95–98. https://doi.org/10.1016/j.enmm.2016.08.002

[22] F. Ijaz, S. Shahid, S.A. Khan, W. Ahmad, S. Zaman, Green synthesis of copper oxide nanoparticles using abutilon indicum leaf extract: Antimicrobial, antioxidant and photocatalytic dye degradation activities, Trop. J. Pharm. Res. 16 (2017) 743–753. https://doi.org/10.4314/tjpr.v16i4.2

[23] S. Saleem, B. Ahmed, M.S. Khan, M. Al-Shaeri, J. Musarrat, Inhibition of growth and biofilm formation of clinical bacterial isolates by NiO nanoparticles synthesized from Eucalyptus globulus plants, Microb. Pathog. 111 (2017) 375–387. https://doi.org/10.1016/j.micpath.2017.09.019

[24] B. Ahmmad, K. Leonard, M. Shariful Islam, J. Kurawaki, M. Muruganandham, T. Ohkubo, Y. Kuroda, Green synthesis of mesoporous hematite (α-Fe2O 3) nanoparticles and their photocatalytic activity, Adv. Powder Technol. 24 (2013) 160–167. https://doi.org/10.1016/j.apt.2012.04.005

[25] N. Thovhogi, E. Park, E. Manikandan, M. Maaza, A. Gurib-Fakim, Physical properties of CdO nanoparticles synthesized by green chemistry via Hibiscus

Sabdariffa flower extract, J. Alloys Compd. 655 (2016) 314–320.
https://doi.org/10.1016/j.jallcom.2015.09.063

[26]   S.S. Shankar, A. Ahmad, M. Sastry, Geranium Leaf Assisted Biosynthesis of Silver Nanoparticles, Biotechnol. Prog. 19 (2003) 1627–1631.
https://doi.org/10.1021/bp034070w

[27]   F.T. Thema, E. Manikandan, M.S. Dhlamini, M. Maaza, Green synthesis of ZnO nanoparticles via Agathosma betulina natural extract, Mater. Lett. 161 (2015) 124–127. https://doi.org/10.1016/j.matlet.2015.08.052

[28]   P.N. Kumar, K. Sakthivel, V. Balasubramanian, Microwave assisted biosynthesis of rice shaped ZnO nanoparticles using Amorphophallus konjac tuber extract and its application in dye sensitized solar cells, Mater. Sci. Pol. 35 (2017) 111–119.
https://doi.org/10.1515/msp-2017-0029

[29]   A.S. Prasad, Iron oxide nanoparticles synthesized by controlled bio-precipitation using leaf extract of Garlic Vine (Mansoa alliacea), Mater. Sci. Semicond. Process. 53 (2016) 79–83. https://doi.org/10.1016/j.mssp.2016.06.009

[30]   G. Sharmila, M. Farzana Fathima, S. Haries, S. Geetha, N. Manoj Kumar, C. Muthukumaran, Green synthesis, characterization and antibacterial efficacy of palladium nanoparticles synthesized using Filicium decipiens leaf extract, J. Mol. Struct. 1138 (2017) 35–40. https://doi.org/10.1016/j.molstruc.2017.02.097

[31]   M.S. Geetha, H. Nagabhushana, H.N. Shivananjaiah, Green mediated synthesis and characterization of ZnO nanoparticles using Euphorbia Jatropa latex as reducing agent, J. Sci. Adv. Mater. Devices 1 (2016) 301–310.
https://doi.org/10.1016/j.jsamd.2016.06.015

[32]   S. Naraginti, P.L. Kumari, R.K. Das, A. Sivakumar, S.H. Patil, V.V. Andhalkar, Amelioration of excision wounds by topical application of green synthesized, formulated silver and gold nanoparticles in albino Wistar rats, Mater. Sci. Eng. C 62 (2016) 293–300. https://doi.org/10.1016/j.msec.2016.01.069

[33]   S.H. Gebre, M.G. Sendeku, New frontiers in the biosynthesis of metal oxide nanoparticles and their environmental applications: an overview, SN Appl. Sci. 1 (2019) 928. https://doi.org/10.1007/s42452-019-0931-4

[34]   M. Shah, D. Fawcett, S. Sharma, S.K. Tripathy, G.E.J. Poinern, Green synthesis of metallic nanoparticles via biological entities, Materials 8 (2015) 7278-7308.
https://doi.org/10.3390/ma8115377

[35]   V.L. Das, R. Thomas, R.T. Varghese, E. V. Soniya, J. Mathew, E.K. Radhakrishnan, Extracellular synthesis of silver nanoparticles by the Bacillus strain CS 11 isolated from industrialized area, 3 Biotech. 4 (2014) 121–126. https://doi.org/10.1007/s13205-013-0130-8

[36]   T.G. Deepak, G.S. Anjusree, S. Thomas, T.A. Arun, S. V. Nair, A. Sreekumaran Nair, A review on materials for light scattering in dye-sensitized solar cells, RSC Adv. 4 (2014) 17615–17638. https://doi.org/10.1039/C4RA01308E

[37]   M.M. Rahman, S.H. Im, J.J. Lee, Enhanced photoresponse in dye-sensitized solar cells via localized surface plasmon resonance through highly stable nickel nanoparticles, Nanoscale 8 (2016) 5884–5891. https://doi.org/10.1039/C5NR08155F

[38]   N. Kulkarni, U. Muddapur, Biosynthesis of metal nanoparticles: A review, J. Nanotechnol. 2014 (2014) Article ID 510246. https://doi.org/10.1155/2014/510246

[39]   G. Gahlawat, A.R. Choudhury, A review on the biosynthesis of metal and metal salt nanoparticles by microbes, RSC Adv., 9 (2019) 12944-12967. https://doi.org/10.1039/C8RA10483B

[40]   J. Huang, L. Lin, D. Sun, H. Chen, D. Yang, Q. Li, Bio-inspired synthesis of metal nanomaterials and applications, Chem. Soc. Rev. 44 (2015) 6330–6374. https://doi.org/10.1039/C5CS00133A

[41]   M. Shafiq, S. Anjum, C. Hano, I. Anjum, B.H. Abbasi, An overview of the applications of nanomaterials and nanodevices in the food industry, Foods. 9 (2020) 1–27. https://doi.org/10.3390/foods9020148

[42]   A. Fakharuddin, R. Jose, T.M. Brown, F. Fabregat-Santiago, J. Bisquert, A perspective on the production of dye-sensitized solar modules, Energy Environ. Sci. 7 (2014) 3952–3981. https://doi.org/10.1039/C4EE01724B

[43]   A. Yella, H.-W. Lee, H.N. Tsao, C. Yi, A.K. Chandiran, M.K. Nazeeruddin, E.W.-G. Diau, C.-Y. Yeh, S.M. Zakeeruddin, M. Gratzel, Porphyrin-Sensitized Solar Cells with Cobalt (II/III)-Based Redox Electrolyte Exceed 12 Percent Efficiency, Science 334 (2011) 629–634. https://doi.org/10.1126/science.1209688

[44]   G. Calogero, A. Bartolotta, G. Di Marco, A. Di Carlo, F. Bonaccorso, Vegetable-based dye-sensitized solar cells, Chem. Soc. Rev. 44 (2015) 3244–3294. https://doi.org/10.1039/C4CS00309H

[45]   A. Hagfeldt, M. Graetzel, Light-Induced Redox Reactions in Nanocrystalline Systems, Chem. Rev. 95 (1995) 49–68. https://doi.org/10.1021/cr00033a003

[46]   K. Prasad, A.K. Jha, ZnO nanoparticles: synthesis and adsorption study, Nat. Sci. 1 (2009) 129–135. https://doi.org/10.4236/ns.2009.12016

[47]   H.A. Atwater, A. Polman, Plasmonics for improved photovoltaic devices, Nat. Mater. 9 (2010) 205–213. https://doi.org/10.1038/nmat2629

[48]   A.K. Mittal, Y. Chisti, U.C. Banerjee, Synthesis of silver nanoparticles plant extracts, Biotech Adv. 31 (2013) 346-356. https://doi.org/10.1016/j.biotechadv.2013.01.003

[49]   K. Ishikawa, C.J. Wen, K. Yamada, T. Okubo, The photocurrent of dye-sensitized solar cells enhanced by the surface plasmon resonance, J. Chem. Eng. Jpn. 37 (2004) 645–649. https://doi.org/10.1252/jcej.37.645

[50]   S. Saravanan, R. Kato, M. Balamurugan, S. Kaushik, T. Soga, Efficiency improvement in dye sensitized solar cells by the plasmonic effect of green synthesized silver nanoparticles, Journal of Science: Advanced Materials and Devices 2 (2017) 418-424. https://doi.org/10.1016/j.jsamd.2017.10.004

[51]   S.C.G.K. Daniel, N. Mahalakshmi, J. Sandhiya, K. Nehru, M. Sivakumar, Rapid synthesis of Ag nanoparticles using Henna extract for the fabrication of Photoabsorption Enhanced Dye Sensitized Solar Cell (PEDSSC), Advanced Materials Research 678 (2013) 349-360. https://doi.org/10.4028/www.scientific.net/AMR.678.349

[52]   C.F. Bohren, D.R. Huffman, Absorption and Scattering of Light by Small Particles (Wiley, New York, 1983).

[53]   C. Nahm, H. Choi, J. Kim, D.-R. Jung, C. Kim, J. Moon, B. Lee, B. Park, The effects of 100 nm-diameter Au nanoparticles on dye-sensitized solar cells, Appl. Phys. Lett. 99 (2011) 253107. https://doi.org/10.1063/1.3671087

[54]   T. Solaiyammal, P. Murugakoothan, Green synthesis of Au and the impact of Au on the efficiency of TiO_2 based dye sensitized solar cell, Materials Science for Energy Technologies 2 (2019) 171–180. https://doi.org/10.1016/j.mset.2019.01.001

[55]   M.N. Asghar, A. Anwar, H.M.A. Rahman, S. Shahid, I. Nadeem, Green synthesis and characterization of metal ions-mixed titania for application in dye-sensitized solar cells, Toxicological & Environmental Chemistry 100 (2018) 659-676. https://doi.org/10.1080/02772248.2019.1590582

[56]   S. Surya, R. Thangamuthu, SM. Senthil Kumar, G. Murugadoss, Synthesis and study of photovoltaic performance on various photoelectrode materials for DSSCs:

Optimization of compact layer on nanometer thickness, Superlattices and Microstructures 102 (2017) 424-441. https://doi.org/10.1016/j.spmi.2017.01.003

[57] S.U. Ekar, G. Shekhar, Y.B. Khollam, P.N. Wani, S.R. Jadkar, M. Naushad, M.G. Chaskar, S.S. Jadhav, A. Fadel, V.V. Jadhav, J.H. Shendkar, R.S. Mane, Green synthesis and dye-sensitized solar cell application of rutile and anatase TiO_2 nanorods, Journal of Solid-State Electrochemistry 21 (2017) 2713–2718. https://doi.org/10.1007/s10008-016-3376-3

[58] I.C. Maurya, S. Singh, S. Senapati, P. Srivastava, L. Bahadur, Green synthesis of TiO_2 nanoparticles using Bixa orellana seed extract and its application for solar cells, Solar Energy 194 (2019) 952–958. https://doi.org/10.1016/j.solener.2019.10.090

[59] R. Rathnasamy, P. Thangasamy, R. Thangamuthu, S. Sampath, V. Alagan, Green synthesis of ZnO nanoparticles using Carica papaya leaf extracts for photocatalytic and photovoltaic applications, J. Mater. Sci.: Mater. Electron. 28 (2017) 10374–10381. https://doi.org/10.1007/s10854-017-6807-8

[60] J.K. Sharma, M.S. Akhtar, S. Ameen, P. Srivastava, G. Singh, Green synthesis of CuO nanoparticles with leaf extract of Calotropis gigantea and its dye-sensitized solar cells applications, J. Alloys and Comp. 632 (2015) 321-325. https://doi.org/10.1016/j.jallcom.2015.01.172

[61] J.K. Sharma, P. Srivastava, G. Singh, M.S. Akhtar, S. Ameen, Green synthesis of Co_3O_4 nanoparticles and their applications in thermal decomposition of ammonium perchlorate and dye-sensitized solar cells, Mater. Sci. and Eng. B 193 (2015) 181–188. https://doi.org/10.1016/j.mseb.2014.12.012

[62] M. Wu, X. Lin, A. Hagfeldt, T. Ma, A novel catalyst of WO_2 nanorod for the counter electrode of dye sensitized solar cells, Chem. Commun. 47 (2011) 4535-4537. https://doi.org/10.1039/c1cc10638d

[63] Y. Liu, S. Yun, X. Zhou, Y. Hou, T. Zhang, J. Li, A. Hagfeldt, Intrinsic origin of superior catalytic properties of tungsten-based catalysts in dye-sensitized solar cells, Electrochim Acta 242 (2017) 390-399. https://doi.org/10.1016/j.electacta.2017.04.176

[64] J. Li, S. Yun, F. Han, Y. Si, A. Arshad, Y. Zhang, B. Chidambaram, N. Zafar, X. Qiao, Biomass-derived carbon boosted catalytic properties of tungsten-based nanohybrids for accelerating the triiodide reduction in dye-sensitized solar cells, J. Colloid and Interface Sci. 578 (2020) 184-194. https://doi.org/10.1016/j.jcis.2020.04.089

[65]   P. Dauthal, M. Mukhopadhyay, Noble Metal Nanoparticles: Plant-Mediated Synthesis, Mechanistic Aspects of Synthesis, and Applications, Ind. Eng. Chem. Res. 55 (2016) 9557-9577. https://doi.org/10.1021/acs.iecr.6b00861

[66]   X. Ji, W. Liu, Y. Leng, A. Wang. Facile synthesis of ZnO@ TiO$_2$ core-shell nanorod thin films for dye-sensitized solar cells, Journal of Nanomaterials 2015 (2015) Article ID 647089. https://doi.org/10.1155/2015/647089

[67]   N. Órdenes-Aenishanslins, G. Anziani-Ostuni, C.P. Quezada, R. Espinoza-González, D. Bravo, J.M. Perez-Donoso, Biological synthesis of CdS/CdSe core/shell nanoparticles and its application in quantum dot sensitized solar cells, Frontiers in microbiology 10 (2019) 1587. https://doi.org/10.3389/fmicb.2019.01587

[68]   B. Gao, C. Shen, B. Zhang, M. Zhang, S. Yuan, Y. Yang, G. Chen, Green synthesis of highly efficient CdSe quantum dots for quantum-dots-sensitized solar cells, Journal of Applied Physics 115 (2014) 193104. https://doi.org/10.1063/1.4876118

[69]   R. Ye, C. Xiang, J. Lin, Z. Peng, K. Huang, Z. Yan, N.P. Cook, E.L.G. Samuel, C.-C. Hwang, G. Ruan, G. Ceriotti, A.-R.O. Raji, A.A. Marti, J.M. Tour, Coal as an abundant source of graphene quantum dots, Nat. Commun. 4 (2013) 2943. https://doi.org/10.1038/ncomms3943

[70]   X. Guo, H. Zhang, H. Sun, M.O. Tade, S. Wang, Green synthesis of carbon quantum dots for sensitized solar cells, ChemPhotoChem 1 (2017) 116–119. https://doi.org/10.1002/cptc.201600038

Bioinspired Nanomaterials for Energy and Environmental Applications   Materials Research Forum LLC
Materials Research Foundations **121** (2022) 211-238   https://doi.org/10.21741/9781644901830-7

Chapter 7

Bioinspired Synthesis of Nanomaterials for Photoelectrochemical Applications

Aruna Kumari M.L.

Department of Chemistry, The Oxford College of Science, Bangalore-560102, India

drarunakumariml@gmail.com

Abstract

Hydrogen production by water splitting using renewable resources like water and solar energy by photoelectrochemical (PEC) processes using biological resources is an endless fountain of inspiration for fabrication and design of nanomaterials/electrodes. In this chapter, we tried to explain how researchers utilized bioinspired materials in creation of new bioinspired synthetic strategies, their artificial mimics and bioinspired process in design of artificial photosynthesis and artificial leaf using PEC technology. This chapter also covers some confront and perspectives in this emerging area of research.

Keywords

Bioinspired Materials, Photoelectrochemical Cell, Water Splitting, Biomimics, Artificial Photosynthesis, Hydrogenases

Contents

Bioinspired Synthesis of Nanomaterials for
Photoelectrochemical Applications 211

1.   Introduction...212

2.   Basic principle of photo electrochemical cell.....................................213

3.   Challenges...214

4.   Bioinspired strategies 216

Conclusion...232

References 232

1.   Introduction

Due to the population growth and a rise in living standards, global energy demand increasing rapidly. Fossil fuels are inadequate to meet future generation's energy needs and another issue is environmental damage caused during the process, transfer, and utilization. Numerous alternative ways of sustainable energy generation methods such as hydropower plants, nuclear power plants, wind turbines, solar, biomass, etc. are used to overcome the energy crisis still, all of them having, respective, drawbacks [1]. The greatest challenge for scientific community has to be addressed is the huge energy requirements and global warming. The development of a novel method using renewable resources such as sunlight and water is a promising solution to the aforementioned problem. In recent years, substantial efforts from both academia and industry have been made towards water splitting into oxygen and hydrogen [2]. From solar-driven water splitting process, solar energy can be converted into chemical energy and it will be stored within the hydrogen molecules. As per demand the stored energy will be converted into electricity by oxidation of hydrogen fuel, which releases water molecules as a byproduct. The important criteria for ideal fuel are cleanliness, inexhaustibility, and convenience. H_2 possesses all these properties and emerges as an environment friendly substitute compared to fossil fuels.

Despite the direct utilization of sunlight for water splitting into hydrogen and oxygen for hydrogen production is futile. Although, UV and visible light photons carry sufficient energy to split water thermodynamically, a direct photolytic cleavage of O–H bonds is not a feasible process because water wouldn't absorb sunlight. Thus, it is mandatory to adopt a photoelectrochemical process that combines light harvesting material, which absorbs the photons and catalyzes the water splitting process. Technological application of this principle leads to the development of the Photo Electrochemical Cell.

Since the discovery of PEC water splitting by Fujishima and Hondain 1972 using TiO_2 photoanode, a huge number of experiments have been conducted and articles are published, this reflects scientist's attempts to resolve the challenges in this field [3]. For practical use, PEC based water splitting reliable photoelectrode materials are required with following criteria: a broad range of solar spectrum harvesting, effective use of charge carriers, reduced high stability for long-term operation, overpotential, and low cost. Even after four decades of extensive research, a visible light driven, efficient and stable water splitting system still remains an elusive goal [4].

Scientists are always seeking inspiration from nature to remove the obstacles. Photosynthesis process act as a fascinating inspiration source to achieve efficient light-to-hydrogen production based on the advantages of PEC technology. From the past decade, the design of biomimetic systems for artificial photosynthesis and the development of bioinspired materials and methods is extensively used as an efficient and effective strategy. The bioinspired approach doesn't rely on imitating or replicating of exact details of an active site, rather it attempts to exploit its basic chemical principles and exciting reactions. The discovery of novel bioinspired materials and processes has achieved many unprecedented results in PEC technology by utilizing the advantages of the natural raw materials directly or indirectly [5].

In this chapter, we first describe the basic principles involved in PEC, and then we discuss challenges and bioinspired strategies employed in photoelectrochemical technology for efficient H₂production.

2. Basic principle of photo electrochemical cell

In PEC systems, an integrated semiconductor/electrolyte junction converts light energy to chemical energy by photo-electrochemical water splitting. The experimental setup consists of a photocatalytic semiconductor as a working electrode, calomel as a reference electrode and counter electrode (usually Pt), and an appropriate supporting electrolyte (Fig. 1). H₂ generation from PEC technology involves various chemical processes within the photoelectrode as well as at photo-electrode and electrolyte interface i.e, on exposure to a light source, semiconducting material undergoes an intrinsic ionization and results in the formation of electron and hole charge carriers Eq. 1; At the anode surface; photoexcited holes oxides the water and drive the oxygen evolution reaction (OER) (Eq. 2); then transport of H^+ ions from the photoanode to cathode takes place via electrolyte and transport of electrons from photoanode to the cathode via external circuit; while photoexcited electron reduces the hydrogen ion (Eq. 3) and leads to hydrogen evolution reaction (HER)and the overall conversion process is shown in Eq. 4[6-9].

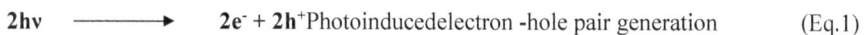

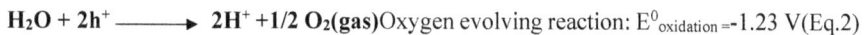

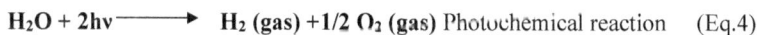

$$2h\nu \longrightarrow 2e^- + 2h^+ \text{ Photoinduced electron -hole pair generation} \quad (Eq.1)$$

$$H_2O + 2h^+ \longrightarrow 2H^+ + 1/2\ O_2(gas) \text{ Oxygen evolving reaction: } E^0_{oxidation} = -1.23\ V (Eq.2)$$

$$2H^+ + 2e^- \longrightarrow H_2(gas) \text{ Hydrogen evolving reaction: } E^0_{reduction} = 0.00\ V \quad (Eq.3)$$

$$H_2O + 2h\nu \longrightarrow H_2\ (gas) + 1/2\ O_2\ (gas) \text{ Photochemical reaction} \quad (Eq.4)$$

Bioinspired Nanomaterials for Energy and Environmental Applications   Materials Research Forum LLC
Materials Research Foundations **121** (2022) 211-238   https://doi.org/10.21741/9781644901830-7

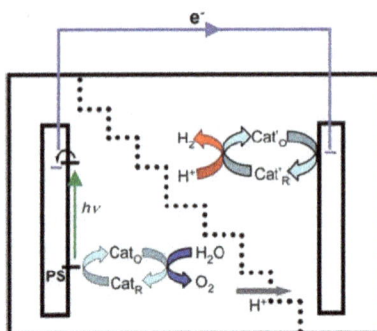

Figure 1. Pictorial representation of photoelectrochemical cell used for water splitting. Reproduced permission from Ref. No. [9]

Alternatively, a two-electrode cell can be connected with photoanode and photocathode without employing a counter electrode. In this tandem configuration, without applying an external bias overall water splitting reaction will take place under light irradiation, in which oxygen and hydrogen evolution take place at the photoanode and photocathode respectively [10-12]. Compared to the regular PEC, where a single semiconductor photoelectrode is used, the tandem PEC cell achieved better performance by pairing up of two semiconductors with required material properties and offers possible synergistic effect. Alternatively when PEC cell coupled with a photovoltaic cell will also result in the unbiased photochemical water splitting, provided that when sufficient photovoltage is contributed by the photovoltaic compartment to make up for the potential deficiency of the PEC cell [13-15]. Moreover, multijunction semiconductor architectures can also be integrated with an electrode made up of electrochemical catalyst, which is also capable of producing desired water splitting without bias. This type of self-powered, unassisted PEC system is the ultimate goal [16-18].

3. Challenges

The conversion of light energy to chemical energy still faces many challenges. The materials used in photoelectrode should satisfy several specific requirements with respect to electrochemical and semiconducting properties and its stability plays a significant role in any photoelectrochemical system. In standard conditions, the transformation of one molecule of H_2O into H_2 and $\frac{1}{2}$ O_2 requires a free energy change of $\Delta G=237.2$ kJ/mol, i.e., according to the Nernst equation, $\Delta E=1.23$ V per transported electron. The basic requirement for a single semiconductor to acts as a photoelectrocatalyst is it must absorb photons with energy greater than 1.23 eV but due to thermodynamic energy losses, a minimum bandgap of 1.8 eV is required for the generation of H_2 and O_2 from water splitting [19].

In two-electrode PEC systems, a semiconductor with either n or p-type conductivity should act as a working electrode. In addition to working and counter electrodes, most of the PEC systems will have reference electrodes to examine half-reactions in the cell. Once the semiconductor in working electrode is illuminated with photons of equal and higher energy level than its bandgap (E_g) create electron-hole pairs, the optimal band gap for high performance photoelectrode is ~2 eV. When working electrode is of n-type semiconductor, the holes oxidize water into O_2 and H^+ and electrons are transferred to the counter electrode and reduce H^+ into H_2. On the other hand, working electrode is p-type semiconductor then, generated electrons are used to H^+ into H_2 and the holes accept the electrons transferred from the counter electrode and leads water oxidization into O_2 and H^+ [20-21].

An ideal photoelectrode material must satisfy following criteria for efficient water splitting. (1) Should possess a strong (visible) light absorption with E_g varying between 1.8 and 2.4 eV (2) Under darkness and illumination it should exhibit chemical inertness and should be of low cost. (3) Should have suitable band edge energy position to support the water redox potentials (4) Should effectively separate or transport the generated charge carriers to enhance the water-splitting reaction rate (5) Should exhibit a low charge transfer resistance (low overpotentials) at the liquid/semiconductor interface.

Accordingly, another major problem should be addressed in PEC development is overpotential. The overpotential is defined as the difference between the potential need to be applied to the system to carry out its function at a significant rate and the standard potential of the redox couple. The determination of overvoltage is easier when the reaction is catalyzed at the electrode surface in aqueous solution but it is difficult to calculate when the catalyst is in the bulk and in non-aqueous solvents. For a process to be economically viable overvoltages should remain below 0.1 V.

Another important aspect in photoelectrode charge transfer dynamics during the water splitting is the reduction of overpotential in presence of cocatalysts. When cocatalysts is introduced it mediates the charge transfer dynamics at the photoelectrode surface and improves surface reaction kinetics For efficient water splitting, the charges need to be balanced in water oxidation and water reduction reactions. Because water oxidation reaction is more difficult and slower than water reduction since it involves a complex four-proton coupled electron transfer process. This highlights the importance of cocatalysts in addition to photoanodes to improve the kinetics of OER. A similar concept can be applied to promote HER kinetics by modifying photocathodes with specific cocatalysts. Despite so much research on the usage of cocatalyst in PEC system, the exact role of it improving the surface reaction kinetics is still a source of debate. This is because the presence of cocatalysts on a photoelectrode surface not only alters the charge transfer kinetics but also changes the chemical environment and electronic structure. Therefore, a detailed study is imperative in the cocatalyst mechanism for the practical implementation of this approach.

Despite of significant progress over the years, the obtained results are not efficient for practical applications [22]. The major three obstacles that exist currently for the practical PEC processes are;

1. The requirement of about 0.6 V above the thermodynamic minimum potential overpotential(which is typically 1.23 V).

2. The generation of H_2O_2 as a byproduct, which poisons and shortens the lifetime of electrodes, as well reduces the anodic current.

3. The third obstacle is the generation of relatively low current densities.

4.   Bioinspired strategies

In the specific field of PEC water splitting, bio-inspired nanomaterials synthesis and bio mimicking structures have out-turned many pioneering results. The significant efforts devoted to this area are generalized in the following categories.

(1) Bio-inspired synthetic strategies: The photoelectrode nanomaterials developed from conventional methodologies are usually suffer from low efficiency and high cost. Hence researchers now utilizing the bio-templates such as bacteria, viruses, porphyrins, proteins, and DNA. These materials remarkably enhance the photoelectrochemical performances at the nanoscale or atomic scale. As we know that small-sized and high crystalline material has a great impact on photoelectrode materials. To make highly crystalline materials, high-temperature calcination is required, this leads to an increase in particle size, thereby decrease in catalytic activity. This can be avoided using bio-template assisted synthesis by carrying out reactions in an aqueous medium in presence of ethylene glycol ligand to bind metal cations present in a semiconducting material; this kind of synthesis takes place exclusively on the virus. There are few related works reported on bacteriophage/virus bio-templates.

D. Oh et al. [23] used M13 filamentous bacteriophage as a template for producing functional nanomaterials upon which, iridium oxide (IrO_2) catalyst and zinc porphyrin (ZnDPEG) photosensitizer are assembled (Fig.2).The protein coat helped to nucleate a

Bioinspired Nanomaterials for Energy and Environmental Applications   Materials Research Forum LLC
Materials Research Foundations **121** (2022) 211-238   https://doi.org/10.21741/9781644901830-7

catalyst, IrO_2, through the peptide motif.6 and controls the thickness of the catalyst. The nanowire network consisting of the ZnDPEG and IrO_2 were precisely spaced to trigger the water splitting reaction; the pigment acts as an antenna to capture the light and transfer the energy along the nanowires, where the catalyst promotes the water-splitting reaction with high quantum efficiency and turnover rate of oxygen production nearly fourfold.

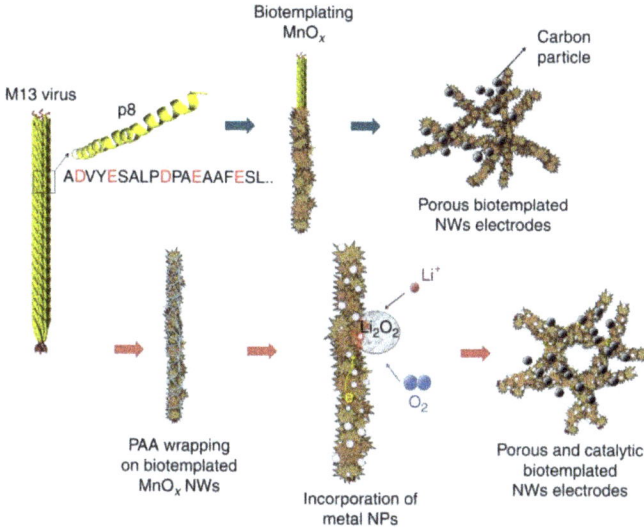

Figure 2 Schematic representation of biotemplated assembly nanowire electrode with photocatalyst.Reproduced permission from Ref. No. [23]

N. Nuraje et al. [24] used genetically engineered M13 virus as a biotemplates on perovskite nanomaterials like strontium titanate (STO) and bismuth ferrite (BFO). The filamentous body of M13 virus comprised of 2700 identical copies of pVIII coat protein. As shown in Fig. 3, the N-terminus of each pVIII possess a high charge density to interact with cationic metal and forms the template. Thus synthesized Virus templated STO nanowires are efficiently used as photoelectrodes under both UV and visible light for H_2production.

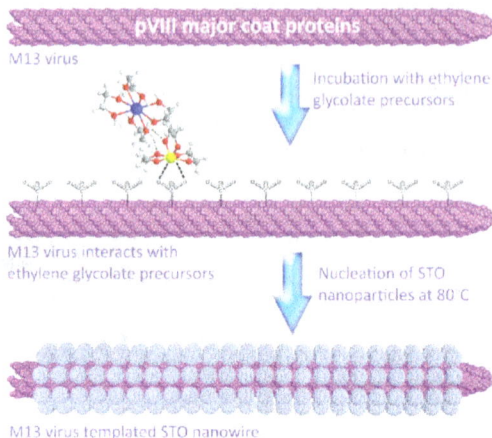

Figure 3. Schematic representation of synthesis M13 Virus-templated STO
nanowire.Reproduced permission from Ref. No. [24]

C. Jolley et al. [25] prepared Cowpea chlorotic mottle virus (CCMV)-TiO$_2$ composite. Photocurrents generated from above composites were collected at an ITO working electrode which is immersed in ananatase–CCMV suspension consists of 0.3 mg mL^{-1}anatase-CCMV, and 20% ethanol in presence of 2 mM of methyl viologen electron mediator. Photo-reactions were performed with the anatase–CCMV composite results photocurrents of the order of 25 mA cm^{-2} and1 mA cm^{-2}when exposed to UV light and 420 nm LED respectively. Conversely, control reactions involving the empty CCMV cage and methyl viologen exhibited photocurrents in the order of 0.0–0.5 mA. Here, the photocurrent observed is a resultant of protein cage oxidation rather than ethanol.

S. Ehrman et al. [26] prepared a 3D patterned current collectors using genetically modified Tobacco mosaic virus (TMV1cys) on gold coated substrates followed by ITO sputtering deposition. Furthermore, sputter deposition of photoactive cupric oxide (CuO) with thicknesses of 520 nm on TMV1cys patterned current collectors produced the highest photocurrent density of 3.15 mA/cm^{2}in comparison with similar-sized CuO system. The combined effects of arises from virus-templated surfaces like increased surface area, suppression of light reflection, and reducing charge carrier transport distance makes it ideal for photoelectrochemical applications. The strategy of combining TMV surface assembly and sputter coating processes is an attractive way for efficientH$_2$ generation.

2)Bio-function Mimicking: The concept of bioinspiration not only attribute to bio-inspired structure but for the biofunctional process. Natural photosynthesis has emerged as a fascinating source of inspiration for artificial photosynthesis systems in designing PEC technology from which efficient H$_2$production will be achieved. Both naturally occurring photosynthesis processes and PEC systems use solar energy however, they operate in different ways. Before proceeding to artificial photosynthesis let's understand water splitting in natural photosynthesis.

Photosynthesis is a natural nanotechnology system that uses macromolecular machines (photosystems) that catalyze solar energy-driven water splitting. In natural photosynthesis, solar light is utilized to produce biomass, oxygen, and hydrogen [27-28]. The entire photosynthesis process can be divided into three steps:

a. light-harvesting process, where 'antenna' in photosystem absorbs photons, induces the formation of electron-hole pair and local charge separation is achieved in Photosystem I (PS-I) and Photosystem II (PS-II) as shown in Fig 4.

b. Larger extent of spatial charge separation is achieved along the photosynthetic chain via proton-coupled electron transfer between redox cofactors.

c. At enzymatic sites such as Fe-Fe cluster in Hydrogenase and CaMnO$_4$ center generation of H$_2$ and O$_2$ takes place respectively through multi-electronic redox catalysis.

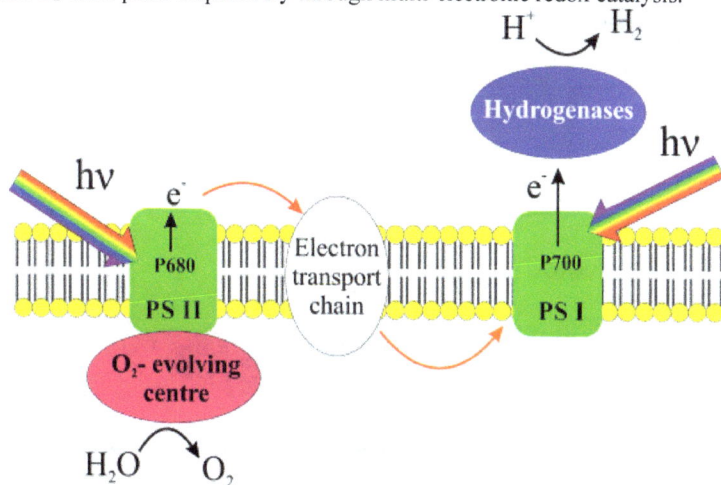

Figure 4: Schematic representation of natural photosynthesis process. Reproduced with permission from Ref. No. [27]

Inspired by Z-scheme photosynthesis, (Fig 5) Y. Che et al. [29] carried out HER using Ag_2CrO_4 decorated g-C_3N_4nanosheetin presence of methanol as a sacrificial agent. The experiment results indicate that the recombination of photogenerated electron-hole pairs is inhibited by Z-scheme g-C_3N_4/Ag_2CrO_4nanosheets, and H_2generation is promoted by water splitting. The nanocomposites g-C_3N_4/Ag_2CrO_4(23.1%) show the 14 times efficiency in HER compared to that of bare g-C_3N_4.

Catalysts play an important role in accelerating multiple chemical reactions such as HER and OER which takes place on the surface of the light harvesters. The following sections deal with progress in biological functions like artificial photosynthesis process and bioinspired materials that mimic the active site responsible for carrying out OER and HER.

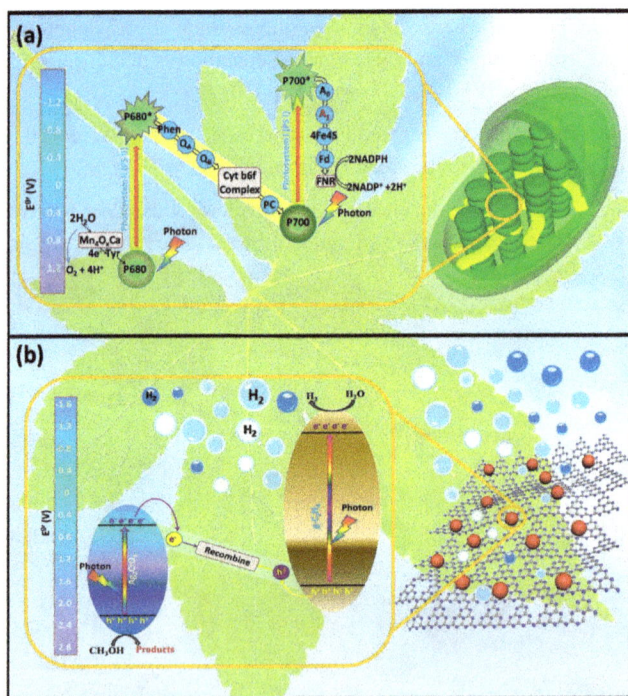

Figure 5 (a) Natural photosynthesis(b) An analogy of artificial photosynthesis composed of g-C_3N_4 (equivalent to PS I) and silver chromate (equivalent to PS II.). Reproduced with permission from Ref. No. [29]

Bioinspired Nanomaterials for Energy and Environmental Applications   Materials Research Forum LLC
Materials Research Foundations **121** (2022) 211-238   https://doi.org/10.21741/9781644901830-7

Oxygen evolution reaction (OER) mimic

Dismukes et al. reported that bioinspired tetra-manganese cluster [$Mn_4O_4L_6$], (L = (*p*-MeOPh)$_2$PO$_2$) supported on a Nafionproton-conducting membrane catalyzes water oxidation under visible light illumination via an intra molecular process at relatively low overpotentials (0.38 V) [30-32]. The water oxidation is achieved by prepared a bio-inspired manganese cubane cluster, where two of the core oxo ligands form an O-O bond resulting in the evolution of O$_2$. They also prepared $Mn_4O_4L_6$cubane like complex with diarylphosphinate ligands which upon photo-oxidation transfer its bridging oxides into O$_2$. Binding of [$Mn_4O_4L_6$]$^+$, to a thin proton-conducting Nafion membrane deposited on a conducting glassy carbon electrode (Fig. 6) yields a photoelectron oxidation of water to O$_2$. A large photocurrent is observed corresponding to the formation of O$_2$when this electrode is illuminated with filtered Xe light (λ > 275 nm) and polarized above the [Mn_4]$^{2+}$/[Mn_4]$^+$ potential at 1.00 V vs Ag/AgCl (electrical bias). Catalytic turnover frequencies were found to be of 20–270 per Mn$_4$ unit and per hour and turnover number is greater than 1000.

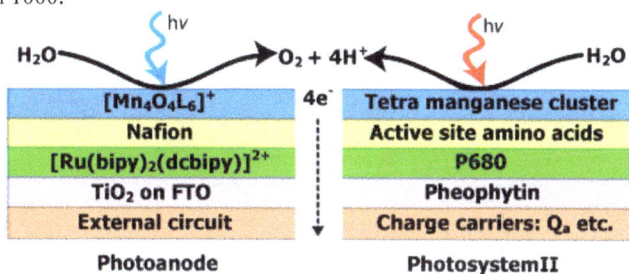

Figure 6. Represents the conceptual similarity of the photoanode Nafion/Ru(4)-TiO$_2$, with photosystem (PSII).Reproduced with permission from Ref. No. [32]

Nocera et al. [33] reported that Cobalt hangman corrole, bearing *meso*-pentafluorophenyl and β-octafluoro substituents act as an active catalyst in water splitting. When this immobilized in Nafion films, the turnover frequencies at single cobalt center of hangman platform approaches 1 s^{-1}for the 4e$^-$/4H$^+$ process. Chorkendorff et al. [34] came up with molybdenum and sulphur based bioinspired molecular clusters with comparable H$_2$ generation rates to that of platinum. The incomplete Mo$_3$S$_4$clusters efficiently catalyze the HER when coupled to a *p*-type Si semiconductor which utilizes red photons of solar spectrum even with no external bias. The current densities at the reversible potential are sufficient for usein 10%solar-to-hydrogen efficiency exhibiting photocurrent density of the order of~ 8mA cm^{-2}.

Harvey J.M. Hou et al. [35] prepared a biomimic substance WO_3 binded Mn-oxophotoanode by atomic layer deposition for efficient water splitting at pH 4 and pH 7. They propose a working model of water splitting using Mn-oxooligmer/WO_3 in Fig. 7. Here, WO_3 acts as the light-harvesting photosystem and generates electrons and holes. The holes in WO_3 are reduced by abstracting electrons from Mn-oxo oligomer via electron transfer reaction. Then Mn-oxo oligomer is oxidized to form an Mn(V) ¼ O species which reacts with H_2O molecules. The O-O bond is formed together with the reduction of Mn(V) ¼ O to a Mn(II)-terpy species and completes the water-splitting cycle. The catalytic cycle of water splitting involves the Mn(II) species and high valent Mn(V)¼O intermediate mimicking the photosynthetic water oxidation..

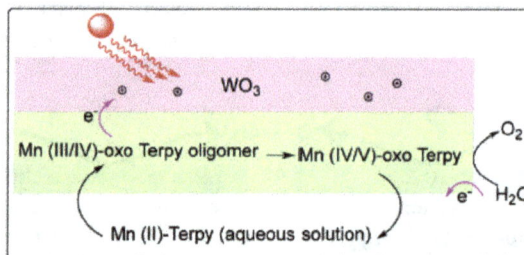

Figure 7shows water splitting process of Mn-oxooligomer complex/WO3 system.Reproduced permission from Ref. No. [35]

R. Nakamura et al. [36] suggested that the efficiency of Mn oxides may be enhanced as OER catalysts by regulating proton coupled electron transfer (PCET) in the electrooxidation step of Mn^{2+} to Mn^{3+}. In PSII, PCET mechanism regulation is a crucial step for the change in redox state of the tetrameric Mn cluster and it is likely promoted by amino acids. However, most bioinspired OER catalysts, particularly Mn oxides, lack such a specific regulatory mechanism under neutral pH conditions. Therefore, they worked on the regulation of PCET with an effective way to minimise the overpotential for water splitting by Mn oxides in neutral media.

C. Li et al. [37] assembled a photoelectron catalytic system as PSII mimick having $BiVO_4$ with layered double hydroxide (NiFeLDH), partially oxidized graphene (pGO),and molecular Co cubane as biomimics of P680, tyrosine and Mn$_4$-Ca cluster respectively. Here

BiVO$_4$ acts as a light harvester and LDH as a hole storage layer, pGO for charge transfer and Co cubane for OER as shown in Fig. 8. The system exhibited low overpotential (0.17 V) and a high photocurrent (4.45 mA cm^{-2}), with a 2.0% solar to hydrogen efficiency. This system acts as biomimetic PEC system.

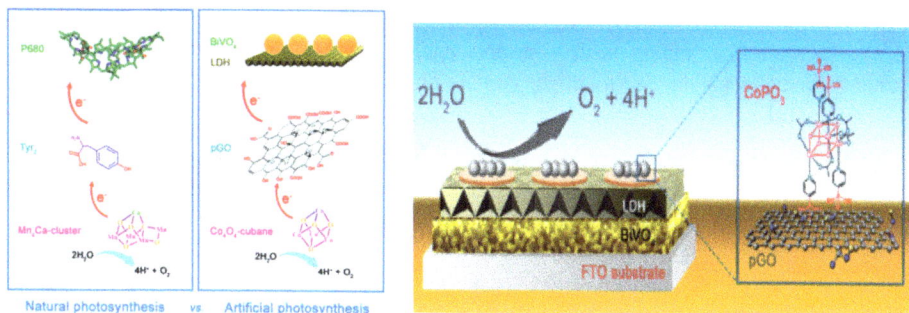

Figure 8. Represents of a) Schematic description of natural and proposed artificial photosynthesis b) Integrated BiVO$_4$Photoanode SystemReproduced permission from Ref. No. [37]

R. Brimblecombe et al. [38] reported photoanode made of a bioinspired cubane cluster, entrapped in a nafion matrix, which is deposited onto a TiO$_2$ anchored ruthenium photosensitizer (Fig. 9). This photoanode under illumination generates photocurrents up to 25 µA cm^{-2}with evolution of O$_2$ when it coupled with platinum counter electrode. This process is very similar to that of PSII. Sun et al. designed a PEC based on the results reported by Brimblecombe [39].

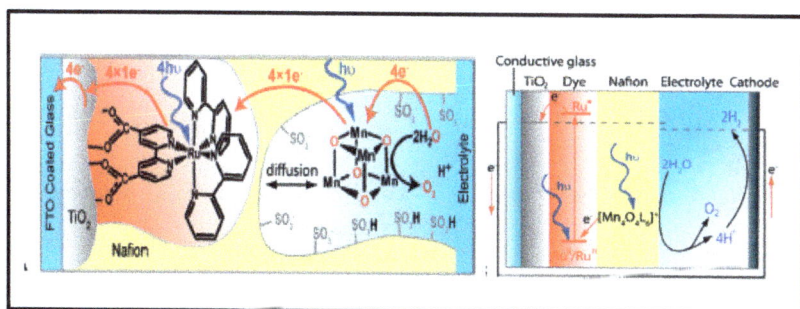

Figure 9 a) Depiction of the photoanode, b) PEC consisting of the photoanode. Reproduced with permission from Ref. No. [38]

Bioinspired Nanomaterials for Energy and Environmental Applications   Materials Research Forum LLC
Materials Research Foundations **121** (2022) 211-238   https://doi.org/10.21741/9781644901830-7

Hydrogen evolution reaction (HER) mimic

Like OER mimics, HER mimics are crucially demanded for H_2 generation. Noble metal-based catalysts are considered as the most effective and robust ones for the HER based application. As we know that Pt catalyses HER with zero overpotential requirement and it also offers high catalytic rate with 0.9 s^{-1}TOF in a pH0 electrolyte solution. New alternatives to the Pt catalyst are in great demand because of its low abundance on earth crust as well as its important uses in other applications. Inspired by biological systems natural metalloenzyme like hydrogenases are discovered as innovative alternatives to Pt catalyst [40].

Hydrogenases are natural enzymes that catalyze the reversible reduction of H^+ to H_2 efficiently. Hydrogenase enzymes exhibit unique properties as HER catalysts, such as low activation energies and a wide scale of O_2 sensitivities. With the use of organometallic sites made up of earth-abundant elements (Fe, Ni, S, C, N, and O) itre-oxidizes the pool of electron carriers in photosynthesis. Hence, hydrogenases are utilized as attractive catalysts, and incorporated into photoactive materials and devices for HER [41]. There are two kinds of hydrogen metabolizing enzymes: the [NiFe] and [FeFe]hydrogenases [9]. The metal site of [NiFe]- and [FeFe]-hydrogenases are shown in Fig. 10.3D structures of hydrogenases reveal the presence of an array of iron-sulfur clusters with less than 15Å distance apart, this facilitates electrical communication between the active site and protein surface where redox partners are expected to bind for accepting (H_2 oxidation) or providing (H_2 formation) electrons [42].

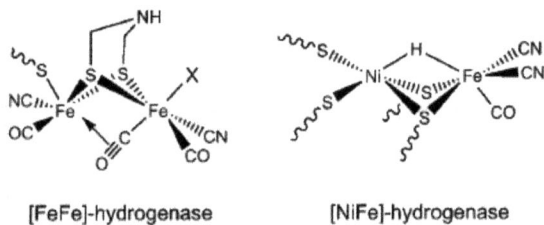

[FeFe]-hydrogenase   [NiFe]-hydrogenase

Figure 10. Active sites of [FeFe]-hydrogenases and [NiFe]-hydrogenases. Reproduced with permission from Ref. No. [9]

Brown et al. [43] have investigated the self-assembly, charge transfer kinetics, and photocatalytic performance of a hybrid complex made up of CdTenano crystals capped with3-mercaptopropionic acid (MPA) and Clostridium acetobutylicum [FeFe]-

hydrogenase I (CaI). As shown in Fig. 11, under visible light irradiation the molecular assembly of MPA capped CdTe and CaIgeneratesH_2 via electrostatic interaction.

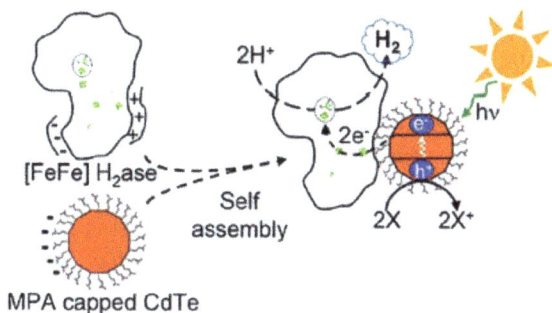

Figure 11Schematic illustration visible light induced H_2production from self-assembly of Clostridium acetobutylicum [FeFe]-hydrogenase I with MPA capped CdTe nanocrystals. Reproduced with permission from Ref. No. [43]

They also found that the CdTe/CaI molar ratio strongly affects the H_2 production. The lower CaI surface coverage results larger contribution of electron transfer to CdTe relaxation [44]. The optimized CdTe–CaI complex achieved in presence of ascorbic acid as a hole scavenger with QE of 9% and 1.8% at 523 nm and under AM 1.5 white light irradiation respectively. ATOF of 25 mol H_2 (molCaI)$^{-1}$ s^{-1} is observed for H_2 production under white light irradiation. In continuation of this work, they employed CaI as an HER cocatalyst by combining it with MPA-capped CdS nanorods. Remarkably, under irradiation at 405 nm the CdS nanorod–CaI complex exhibits a TOF of 380–900 mol H_2(molHydrogenase)$^{-1}$ s^{-1}and a QE of up to 20% and TON of 106 after H_2 evolution for 4 hr. This loss of activity is attributed to oxidative loss of the MPA ligand from the CdS surface by photoexcited holes

Results the deactivation of CaI. The results not only demonstrate the synergy between coupling semiconductor photocatalysts and biological hydrogenase cocatalysts in H_2 production but also provide an insight into the rational design and preparation of the hybrid hydrogenase and semiconductor photocatalyst in the future.

Many research groups are engaged in developing small organometallic catalysts as a mimic of hydrogenase enzymes and these artificial catalysts are named hydrogenase mimics. In the past two decades, several hydrogenase mimics have been developed.

Ihara et al. [45] first reported an example of solar-driven H_2 production system using hydrogenase and PS I. They designed an artificial fusion protein-membrane consists of [NiFe] hydrogenase and PS I subunit. This hydrogenase-PS I complex results H_2 production at a rate of 0.58 µmol of H_2 (mg of Chl)$^{-1}$h^{-1} in presence of electron donor system consists of ascorbic acid, DTT, and TMPD. Heberle et al. [46] developed a membrane-bound hydrogenase from the histidine-tagged form of PS Isubunit and Eutropha H16, which was immobilized on nickel-functionalized gold electrode surface via a nickel His-Tag interaction. Under visible illumination (λmax = 700 nm) a photo current of 85 nA cm^{-2} was observed and 30% of it is utilized for HER.

Golbeck et al. [47] has developed a molecular wire to connect two proteins of relevant redox cofactors for direct electron transfer via light-induced H_2 generation. This is achieved by forming a dithiol link between Fe-S bond in [4Fe-4S] cluster of PS I and [4Fe-4S] cluster of [FeFe]-hydrogenase from C. acetobutylicum. Thus developed nanowire established a direct electron transfer between the two components, this upon illumination results H_2 production at a rate of 30.3 µmol of H_2 (mg of Chl)$^{-1}$ h^{-1}, at pH = 8.3 in the presence of a cytochrome c_6, ascorbate and phenazinemethosulfate as sacrificial electron donor system. This PS I - Hydrogenase system remained stable and active more than 2 months in the absence of oxygen.

E. Reisner et al. [48] reported the semibiological CNx–H_2ase and purely synthetic CNx–NiP hybrid system for visible-light-driven H_2 generation with TONs of more than 50 000 mol H_2 (mol H_2ase)$^{-1}$ and approximately 155 mol H_2 (mol NiP)$^{-1}$ in redox-mediator free aqueous solution at pH 6 and 4.5 respectively. The additional use of methyl viologen as redox mediator in CN$_x$–H_2ase hybrid allowed H_2 generation with TOF of 12.3s^{-1}and a TON of >1X10^6. P. D. Tran et al. [49] reported first operational noble metal-free Proton exchange membrane fuel cell (PEMFC) using a bio-inspired Ni-based material and a Co-based material at the anode and cathode respectively. They synthesized a water oxidation catalyst i.e, nickel bisdiphosphine complex inspired by the active sites of Fe-Fe and Ni-Fe hydrogenases (Fig.12). The bis-diphosphine nickel complexes offers nickel centre electron-rich environment as found in Ni-Fe hydrogenases and aza-propanedithiola to cofactor of Fe-Fe hydrogenases. In both classes of hydrogenases, the activation and heterolytic cleavage of H_2takes place in presence of basic residues at catalytic metal centre.

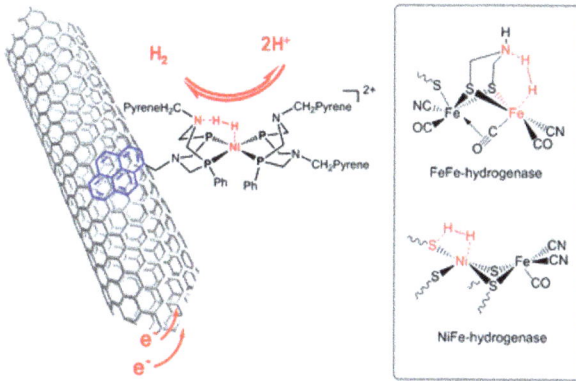

Figure 12 Structure of nickel bisdiphosphine complex anchored to CNT catalyst inspired by Fe-Fe and Ni-Fe hydrogenases shown in the inset. Reproduced with permission from Ref. No. [49]

N. H. Voelcker et al. [50] reported a strategy to fabricate the photoelectrode for a PEC water-splitting from porous silicon nanoparticles (pSi NPs) coupled with indium phosphide nanocrystals and bio-inspired iron sulfur carbonyl ($Fe_2S_2(CO)_6$) catalyst (Fig 13). This is the first example a photocathode for H_2 production made up pSi NP system sensitized with inorganic nanomaterials with a photocurrent density of -2.2 $\mu A/cm^2$ and 5.84 nmol of H_2 per hour.

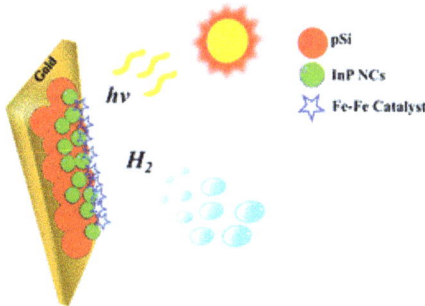

Figure 13 Schematic representation of photoelectrode made up of pSi NPs, InP NCs + $Fe_2S_2(CO)_6$ catalyst. Reproduced with permission from Ref. No. [50]

B. Kumar et al. [51] reported an inexpensive PEC for the homogenous catalytic reduction of H^+ to H_2, which consists of an illuminated and biased p-type Silicon (Si) photocathode, a glassy carbon electrode, and an electrocatalyst [Fe_2 (μ-bdt)$(CO)_6$] designated as [FeFe], with perchloric acid ($HClO_4$)as a proton source. The [FeFe] complex can be reduced reversibly by a single two-electron process at -1.32 V versus Fe^+ /Fe^0, exhibits a high rate constant for proton reduction at approximately 500 $m^{-1}s^1$ in the presence of strong and maintains its structural integrity to a large extent even in the reduced and protonated state. Furthermore, T. Nann et al. [52] constructed a nanophotocathode array for H_2 production by modifying gold electrodes by adsorbing a monolayer of 1,4-benzenedithiol, which helps in binding of a primary layer of indium phosphide (InP)to the gold. Layer-by-layer buildup of the nanoparticle assembly was achieved by alternate exposure to the 1,4-benzenedithiol and the InP solutions. Finally, the [$Fe_2S_2(CO)_6$] cluster which is a mimic of Fe-Fe hydrogenase was introduced into the assembly. As we know the [FeFe]-hydrogenase catalyzes the reduction of H^+ to H_2 under dark conditions at potentials between -0.7 and -1.4 V versus the standard calomel electrode (SCE) in non-aqueous electrolytes. Hence author chose [$Fe_2S_2(CO)_6$] instead of hydrogenase, which has sulfide bridges that are capable of binding to indium as a catalyst for photoelectrochemical reduction of H^+ in a solid-state assembly with modest reduction potential of -0.90 V versus SCE. With this system, they able to achieve more than 60%, photochemical efficiency of which is a breakthrough in this field.

T.A. Moore et al. [53] developed a photoelectrochemical biofuel cell where hydrogenase-based photocathode is developed from pyrolytic graphite edge and carbon felt electrode surfaces modified by adsorbing [FeFe]-hydrogenase from Clostridium acetobutylicum (Fig. 14). And photoanode is made of porphyrin-sensitized TiO_2 nanoparticles supported on conductive F doped tin oxide (FTO) substrate. Armstrong's group utilizes the O_2-resistance property of the [Ni-Fe] hydrogen as enzyme in Ralstonia species, (R. metallidurans and R. eutropha) in H_2 production [54]. The reduction process of H^+ to H_2 is observed at pH 5.5and 40°C under N_2 atmosphere containing, with70 s^{-1}TOF of without any significant overvoltage. This data provides an evidence that a hydrogenase-based electrode can be used practically as an electrocatalytic material in water spilitting. Li et al. [55] developed an artificial photosynthetic hybrid system for H_2 evolution using [Fe-Fe]-H_2ase mimic. They used ZnS as light harvester harvester,[(μ-SPh-4-NH_2)$_2Fe_2(CO)_6$] as a cocatalyst, and ascorbic acid as an electron donor. This system showed a high H_2evolution activity and good stability with TOF of 100 h^{-1} and TON of > 2607 up to 38 h. Wu et al. [56] prepared interface-directed assembly using CdSeQDs and [FeFe]-H_2ase mimic as a water-soluble artificial photosynthetic system. Under visible-light illumination this system showed a very high efficiency for H_2 evolution with a TOF of 596 $h^{-1}$and TON of 8781.

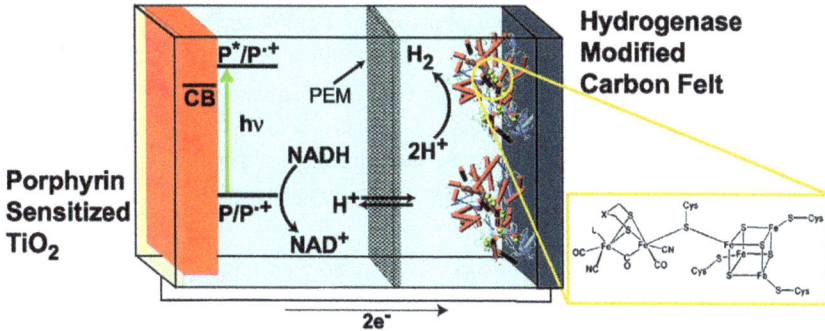

Figure 14 Schematic representations of Moore's photoelectrochemical cells using surface-immobilized hydrogenases as catalyst for H₂ evolution. Reproduced with permission from Ref. No. [53]

Inspired $[Ni^{II}(P^R_2N^{R'}_2)_2]^{2+}$ bio-mimics T. N. Huan et al. [57] synthesized NiPCy$_2$-functionalized on multi-wall carbon nanotubes (MWCNTs) which persists a unique properties for bidirectional and reversible hydrogen evolution. In PEC cells an overpotential is required to reach current density of 10 or 20 mA.cm^{-2} for H$_2$ evolution. However, the NiPCy$_2$-functionalized MWCNT electrode requires only 32 and 60 mV to reach 10 or 20 mA.cm^{-2} current density at 85°C, which is nearly identical to that of the platinum membrane-electrode assembly.

Even though [FeFe]-Hydrogenase mimics exhibits better H$_2$-evolution activity reaction must be carried out in organic solvents or in mixtures of water and organic solvent due to their dissolubility in water, which is neither economic nor environmentally benign. Hence Wang et al. have been put forward to solve this problem [58]. They developed an artificial water-soluble [FeFe]-H$_2$ase mimic by incorporating a cyanide group onto three hydrophilic ether chains to the active site of the [FeFe]-H$_2$ase to improve its solubility in water. The generated the H$_2$ in aqueous solution containing [FeFe]-H$_2$ase mimic which is coupled with CdTe QDs in presence of ascorbic acid as an electron donor. This system produces 786 mmol (17.6 mL) H$_2$ upon the irradiation >400 nm photons with 505 and 50 h^{-1} values TON and TOF respectively even after 10 hours.

(3) Artificial bio-structures: Biological systems are functioning perfectly with maximum performance is owing to their sophisticated hierarchical structure. With this inspiration,

Bioinspired Nanomaterials for Energy and Environmental Applications   Materials Research Forum LLC
Materials Research Foundations **121** (2022) 211-238   https://doi.org/10.21741/9781644901830-7

researchers developed and utilized a wide variety of hierarchical structure in the structural design of electrode materials with optimal properties. Recently, great interest was shown on artificial bio-structures in H_2 production using PEC technology. Among them few are reported here.

F. Peng et al. [59] designed a novel inner-motile film for photocatalytic water splitting. The film is made up of three functional integrates namely ZnO nanowires arrays, CdS quantum dots, and magnetically actuated artificial cilia. These three integrates work synergistically and enhance the photocatalytic H_2 evolution.

Forest-like hierarchical structure is prepared by grafting of ZnO nanowires arrays and CdS on magnetically actuated artificial cilia, this increases the surface area and light absorption. Under illumination, ZnO/CdS heterostructures enhance electron–hole separation and interfacial charge transfer via Z-scheme mechanism, in which ZnO and CdS acts as PS II and PS I, respectively (Fig. 15). Consequently, the rate ofH_2evolution from ZnO/CdS grafted on artificial cilia is about 2.7 times, 2.0 times of ZnO/CdS heterostructures. This film mimics ciliary motion under a rotational magnetic field, exhibiting a singular ability of microfluidic manipulation which resolves the problem of desorption of H_2 and promotes the release of active sites.

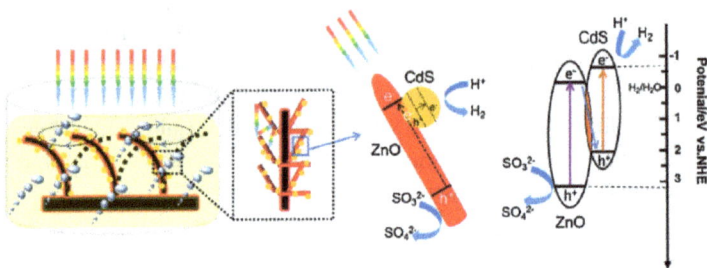

Figure 15.Schematic representation of the charge transfer in the Z-scheme mechanism in the ZnO/CdS grafted on artificial cilia film for hydrogen evolution. Reproduced with permission from Ref. No. [59]

J. Wang et al. [60] prepared a junction hexagonal-cubic phase of $ZnIn_2S_4$ through a bio-inspired approach using hydrilla template (Fig. 16). Under visible-light illumination, this hierarchy enhances photocatalytic H_2 generation in comparison to those single-phase structures hexagonal and cubic. The enhancement in photocatalytic performance is attributed to (i) more efficient charge separation of hexagonal/cubic $ZnIn_2S_4$ heterojunction, (ii) light absorption and harvesting of special hydrilla leaves structure.

.

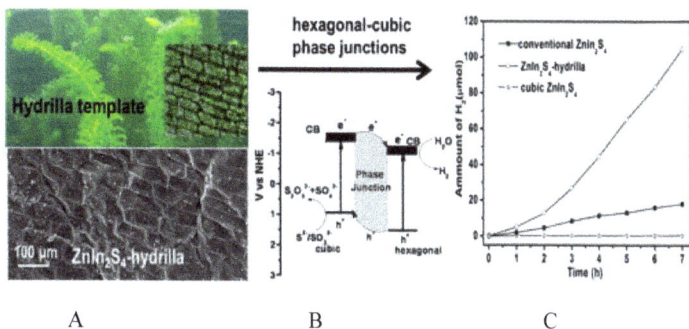

Figure 16.A) SEM image of $ZnIn_2S_4$hydrillaB) Schematicrepresentation of proposed mechanism C) hydrogen evolution. Reproduced with permission from Ref. No. [60]

T. Fan et al. [61] prepared Pt/N-doped TiO_2 artificial inorganic leaves for efficient H_2O splitting under UV and visible light irradiation in the presence of sacrificial reagents. When 2wt% Pt nanoparticles are directly grown onto the artificial TiO_2 leaf and the H_2 evolution rate increases to 1401.70mmol h^{-1} in 20% aqueous methanol, which is about 10 times higher than bare TiO_2 artificial inorganic leaves.

Inspired by superhydrophobic surfaces in nature, electrodes with hydrophobic surfaces submerged in water have attracted much attention of the researchers and various nanomaterials have been developed in this concern. J. Lee et al. [62] attempted to combine two biomimetic properties of lotus natural leaves, i.e., super hydrophobicity and artificial photosynthesis for H_2 generation by solar water splitting. They developed ZnO/Si hierarchical structure for H_2 generation. The rationally designed n/p junction in the ZnO/Si hierarchical structure demonstrated a long plastron lifetime and super hydrophobicity. Nocera et al. [63] have also designed a solar water splitting device in order avoids the use of expensive noble metal catalysts. They replace the earth-abundant Cobalt–Phosphate and a NiMoZn-alloy with noble metal catalysts in water oxidation and H_2production respectively. In their 'artificial leaf', visible light is absorbed by triple-junction silicon deposited onto a stainless steel support and coated with an ITO layer. The device achieved solar-to-hydrogen efficiencies of 4.7% and 2.5% for the wired and wireless configurations respectively. X. Wang et al. [64] synthesized hollow carbon nitride nanosphere inspired by thylakoids. These hollow nanospheres acts as both light-harvesting antennae and nanostructured scaffolds to improve photo redox catalysis with 7.5% apparent quantum yield.

Conclusion

This book chapter reviews the fundamental principles of PEC water splitting and the scientific challenges of visible light driven water splitting and how to overcome those challenges by the inspiration of nature. From biological inspiration research developed various bioinspired synthetic processes and biotemplating provides a new way of controlling the incorporation and precise deposition of catalytic materials in the electrode assembly. While challenges remain concerning the development of high-efficiency, corrosion-resistant photoelectrodes, with exotic physicochemical properties for photoelectrochemical cell development for H_2 generation. Inspired by natural photosynthesis researchers attempted to mimic it by artificial photosynthesis by utilizing OER and HER catalyst and their artificial mimics. The scientific path is steadily heading towards establishing a model sunlight triggered catalytic system for renewable fuel generation using water as raw material. However, there are still many challenges and open questions in the field, regarding both scientific as well as technological aspects but still they are working hard towards demonstration of an efficient practical device for solar fuel a reality.

References

[1] E. Barbier, Geothermal energy technology and current status: an overview. Renew Sustain Energy Rev. 6 (2002) 3-65. https://www.sciencedirect.com/science/article/abs/pii/S1364032102000023

[2] G. Wang, Y. Ling, H. Wang, L. Xihong, Y. Li, Chemically modified nanostructures for photoelectrochemical water splitting, J. Photochem. Photobiol. C: Photochem. Rev. 19 (2014) 35–51. https://www.sciencedirect.com/science/article/pii/S1389556713000439.

[3] A. Fujishima, K. Honda, Electrochemical photolysis of water at a semiconductor electrode. Nature, 238 (1972) 37-38. https://doi.org/10.1038/238037a0

[4] Yi-H. Chiu, T. H. Lai, M-Yu Kuo, P-Y Hsieh, Y- Hsu, Photoelectrochemical cells for solar hydrogen production: Challenges and opportunities,APL Mater. 7, (2019) 080901(1-11). https://aip.scitation.org/doi/10.1063/1.5109785

[5] M. Ding, G. Chen, W. Xu, C. Jia, H. Luo, Bio-inspired synthesis of nanomaterials and smart structures for electrochemical energy storage and conversion, Nano Materials Science, 2, Issue (2020) 264-280. https://doi.org/10.1016/j.nanoms.2019.09.011

[6] M. Gratzel, Photoelectrochemical cells, Nature, 414 (2001) 338-344.https://doi.org/10.1038/35104607

[7] T. Bak, J. Nowotny, M. Rekas, C.C. Sorrell, Photo-electrochemical hydrogen generation from water using solar energy. Materials-related aspects.Int. Jour. of Hyd. Ene., 27 (2002) 991-1022. https://doi.org/10.1016/S0360-3199(02)00022-8

[8] R. Krol, Photoelectrochemical Hydrogen Production, Electronic Materials: Science & Technology, Springer, 2012, https://doi.org/10.1007/978-1-4614-1380-6-2.

[9] P. D. Tran, V. Artero, M. FontecaveWater electrolysis and photoelectrolysis on electrodes engineered using biological and bio-inspired molecular systems, Energy Environ. Sci., 3 (2010) 727–747. https://doi.org/10.1039/B926749B

[10] J.-W. Jang, C. Du, Y. Ye, Y. Lin, X. Yao, J. Thorne, E. Liu, G. McMahon, J. Zhu, A. Javey, J. Guo, D. Wang, Enabling unassisted solar water splitting by iron oxide and silicon, Nat. Commun. 6 (2015) 7447 (1-5). https://doi.org/10.1038/ncomms8447

[11] C. Liu, J. Tang, H. M. Chen, B. Liu, and P. Yang, A Fully Integrated Nanosystem of Semiconductor Nanowires for Direct Solar Water Splitting Nano Lett. 13 (2013) 2989–2992.https://doi.org/10.1021/nl401615t

[12] H. B. Yang, J. Miao, S.-F. Hung, F. Huo, H. M. Chen, B. Liu, Stable Quantum Dot Photoelectrolysis Cell for Unassisted Visible Light Solar Water Splitting ACS Nano 8, (2014) 10403–10413. https://doi.org/10.1021/nn503751s

[13] J. Brillet, J.-H. Yum, M. Cornuz, T. Hisatomi, R. Solarska, J. Augustynski, M. Graetzel, K. Sivula,Highly efficient water splitting by a dual-absorber tandem cell. Nat. Photonics 6, 824–828 (2012).https://doi.org/10.1038/nphoton.2012.265

[14] X. Shi, K. Zhang, K. Shin, M. Ma, J. Kwon, I. T. Choi, J. K. Kim, H. K. Kim, D. H. Wang, J. H. Park,Unassisted photoelectrochemical water splitting beyond 5.7% solar-to-hydrogen conversion efficiency by a wireless monolithic photoanode/dye-sensitised solar cell tandem device, Nano Energy 13(2015)182–191. https://doi.org/10.1016/j.nanoen.2015.02.018

[15] J. H. Kim, Y. Jo, J. H. Kim, J. W. Jang, H. J. Kang, Y. H. Lee, D. S. Kim, Y. Jun, J. S. Lee, Wireless Solar Water Splitting Device with Robust Cobalt-Catalyzed, Dual Doped $BiVO_4$ Photoanode and Perovskite Solar Cell in Tandem: A Dual Absorber Artificial Leaf. ACS Nano 9 (2015)11820–11829. https://doi.org/10.1021/acsnano.5b03859

[16] J. L. Young, M. A. Steiner, H. Doscher, R. M. France, J. A. Turner, T. G. Deutsch, Direct solar-to-hydrogen conversion via inverted metamorphic multi-junction semiconductor architectures. Nat. Energy 2 (2017) 17028(1-8). https://doi.org/10.1038/nenergy.2017.28

[17] E. Verlage, S. Hu, R. Liu, R. J. R. Jones, K. Sun, C. Xiang, N. S. Lewis, H. A. Atwater, A monolithically integrated, intrinsically safe, 10% efficient, solar-driven water-splitting system based on active, stable earth-abundant electrocatalysts in conjunction with tandemIII–V light absorbers protected by amorphous TiO_2 films, Energy Environ. Sci. 8 (2015) 3166–3172. https://doi.org/10.1039/C5EE01786F

[18] M. M. May, H.-J. Lewerenz, D. Lackner, F. Dimroth, T. Hannappel, Efficient direct solar-to-hydrogen conversion by *in situ* interface transformation of a tandem structure,Nat. Commun. 6 (2015) 8286 (1-7).https://doi.org/10.1038/ncomms9286

[19] P. Arunachalam, A. M. Al Mayouf, Chapter 28 Photoelectrochemical Water Splitting, ISBN: 978-0-12-814134-2. https://doi.org/10.1016/B978-0-12-814134-2.00028-0

[20] C. Jiang, S. J. A. Moniz, A. Wang, T. Zhang J. Tang, Photoelectrochemical devices for solar water splitting – materials and challenges,Chem. Soc. Rev. 46 (2017)4645–4660. https://doi.org/10.1039/C6CS00306K

[21] T. G. Vo, J. M. Chiu, C. Y. Chiang and Y. Tai, Solvent-engineering assisted synthesis and characterization of $BiVO_4$ photoanode for boosting the efficiency of photoelectrochemical water splitting, Sol. Energy Mater. Sol. Cells, 166 (2017) 212–221. https://doi.org/10.1016/j.solmat.2017.03.012

[22] W. Zhang, K. Banerjee-Ghosh, F. Tassinari, R. NaamanEnhanced Electrochemical Water Splitting with Chiral Molecule-Coated Fe_3O_4 Nanoparticles, ACS Energy Lett., 3(2018) 2308−2313. https://doi.org/10.1021/acsenergylett.8b01454

[23] D. Oh, J. Qi, Y.-C. Lu, Y. Zhang, Y. Shao-Horn, A. M. Belcher, Biologically enhanced cathode design for improved capacity and cycle life for lithium-oxygen batteries, Nat. Commun., 4 (2013) 2756 (1-8). https://doi.org/10.1038/ncomms3756

[24] N. Nuraje , X. Dang, J. Qi, M. A. Allen,Y. Lei, A. M. Belcher, Biotemplated Synthesis of PerovskiteNanomaterials for Solar Energy Conversion, Adv. Mater. 24 (2012) 2885–2889. https://doi.org/10.1002/adma.201200114

[25] C. Jolley, M. Klem, R. Harrington, J. Parise, T. Douglas, Structure and photochemistry of virus capsid-TiO2nanocomposite, Nanoscale, 3(2011) 1004-1007. https://doi.org/10.1039/C0NR00378F

[26] C-Y. Chiang, J. Epstein, A. Brown, J. N. Munday, J. N. Culver, S. Ehrman, Biological Templates for Antireflective Current Collectors for Photoelectrochemical

Cell Applications,   Nano Lett. 12 (2012) 6005−6011.https://doi.org/10.1021/nl303579z

[27] P. D. Nguyen, T. M. Duong , P. D. TranCurrent progress and challenges in engineering viable artificial leaf for solar water splittingJ. Sci. Adv. Mater.Devices 2 (2017) 399-417.https://doi.org/10.1016/j.jsamd.2017.08.006

[28] E. Andreiadis, M. Chavarot-Kerlidou, M. Fontecave, V. Artero, Artificial Photosynthesis: From Molecular Catalysts for Light-driven Water Splitting to Photoelectrochemical Cells. Photochem.andPhotobio., 87 (2011) 946–964. https://doi.org/10.1111/j.1751-1097.2011.00966.x

[29] Y. Che, B.Lu, Q. Qi, H. Chang, J.Zhai, K. Wang, Z. Liu, Bio-inspired Z-scheme g-C_3N_4/Ag_2CrO_4 for efficient visible light photocatalytic hydrogen generation, Sci. Rep. 8 (2018) 16504 (1-12). https://doi.org/10.1038/s41598-018-34287-w

[30] W. F. Ruettinger, C. Campana, G. C. Dismukes, Synthesis and characterization of $Mn_4O_4L_6$ complexes with cubane-like core structure: A new class of models of the active site of the photosynthetic water oxidase, J. Am. Chem. Soc. , 119 (1997) 6670-6671. https://doi.org/10.1021/ja9639022

[31] R. Brimblecombe, G. F. Swiegers, G. C. Dismukes, L. Spiccia, Sustained Water Oxidation Photocatalysis by a Bioinspired Manganese Cluster, Angew. Chem. Int. Ed., 120 (2008) 7335 -7338. https://doi.org/10.1002/anie.200801132

[32] G. C. Dismukes, R. Brimblecombe, G. A. N. Felton, R.S. Pryadun, J. E. Sheats, L. Spiccia, G. F. Swiegers, Development of Bioinspired Mn_4O_4-Cubane Water Oxidation Catalysts: Lessons from Photosynthesis. Acc. Res. Chem. Res., 42 (2009) 1935-1943. https://doi.org/10.1021/ar900249x

[33] D. Dogutan, R. McGuire, D. G. Nocera, Electocatalytic Water Oxidation by Cobalt(III)Hangman β-OctafluoroCorroles. J. Am. Chem. Soc., 133 (2011) 9178–9180. https://doi.org/10.1021/ja202138m

[34] Y. Hou, B.L. Abrams, P.C.K. Vesborg, M. E. Björketun, K. Herbst, L. Bech, A.M. Setti, C. D. Damsgaard, T. Pedersen, O. Hansen, J. Rossmeisl, S. Dahl, J. K. Nørskov, Ib Chorkendorff, Bioinspired molecular co-catalysts bonded to asilicon photocathode for solar hydrogen evolution. Nat. Mat.,10 (2011) 434-438. https://doi.org/10.1038/nmat3008

[35] L-Y Chou, R. Liu, W. He, N. Geh, Y. Lin, E.Y.F. Hou, D.Wang, H. J.M. Hou, Direct oxygen and hydrogen production by photo water splitting using a robust bioinspired manganese-oxo oligomer complex/tungsten oxide catalytic system,Int J Hydrogen Energy, 37 (2012)8889-8896. https://doi.org/10.1016/j.ijhydene.2012.02.074

[36] A. Yamaguchi, R. Inuzuka, T. Takashima, T. Hayashi, K. Hashimoto, R, Nakamura, Regulating proton-coupled electron transfer for efficient water splitting by manganese oxides at neutral pH.Nat Commun., 5 (2014) 4256 (1-6). https://doi.org/10.1038/ncomms5256

[37] S. Ye, C. Ding, R. Chen, F. Fan, P. Fu, H. Yin, X. Wang, Z. Wang, P. Du, C. Li, Mimicking the Key Functions of Photosystem II in Artificial Photosynthesis for Photoelectrocatalytic Water Splitting , J. Am. Chem. Soc. 140 (2018) 3250–3256. https://doi.org/10.1021/jacs.7b10662

[38] L. Spiccia, R. Brimblecombe, A. Koo, G. Dismukes and G. Swiegers, Solar driven water oxidation by a bioinspired manganese molecular catalyst. J. Am. Chem. Soc., 132 (2010) 2892–2894. https://doi.org/10.1021/ja910055a

[39] L. Li, L.L. Duan, Y.H. Xu, M. Gorlov, A. Hagfeldt, L.C. Sun, A photoelectrochemical device for visible light driven water splitting by a molecular ruthenium catalyst assembled on dye-sensitized nanostructured TiO_2. Chem.Commun., 46 (2010) 7307- 7309. https://doi.org/10.1039/C0CC01828G

[40] T.F. Jaramillo, K.P. Jørgensen, J. Bonde, J.H. Nielsen, S. Horch, I. Chorkendorff, Identification of active edge sites for electrochemical H2 evolution from MoS2nanocatalysts, Science 317 (2007) 100- 102. https://doi.org/10.1126/science.1141483

[41] J. Ran, J. Zhang, J. Yu, M. Jaroniec,S. Z. Qiao, Earth-abundant cocatalysts for semiconductor-based photocatalytic water splitting,Chem. Soc. Rev., 43 (2014) 7787- 7812.https://doi.org/10.1039/C3CS60425J

[42] C. Tard,C. J. Pickett, Structural and Functional Analogues of the Active Sites of the [Fe]-, [NiFe]-, and [FeFe]-Hydrogenases, Chem. Rev., 109 (2009) 2245- 2274.https://doi.org/10.1021/cr800542q

[43] K. A. Brown, S. Dayal, X. Ai, G. Rumbles, P. W. King, Controlled Assembly of Hydrogenase-CdTeNanocrystal Hybrids for Solar Hydrogen Production, J. Am. Chem. Soc., 2010, 132, 9672-9680.https://doi.org/10.1021/ja101031r

[44] K. A. Brown, M. B. Wilker, M. Boehm, G. Dukovic, P. W. King, Characterization of Photochemical Processes for H2 Production by CdSNanorod–[FeFe] Hydrogenase Complexes, J. Am. Chem. Soc., 134 (2012) 5627- 5636.https://doi.org/10.1021/ja2116348

[45] M. Ihara, H. Nishihara, K.S. Yoon, O. Lenz, B. Friedrich, H. Nakamoto, K. Kojima , D. Honma, T, Kamachi, I. Okura, Light-driven hydrogen production by a hybrid complex of a [NiFe]-hydrogenase and the cyanobacterial photosystem I. Photochem. Photobiol,. 82 (2006) 676–682. https://doi.org/10.1562/2006-01-16-RA-778

[46] J. Heberle, H. Krassen, A. Schwarze, B. Friedrich, K. Ataka, O. Lenz, Photosynthetic hydrogen production by a hybrid complex of photosystem I and [NiFe]-hydrogenase. ACS Nano., 3 (2009) 4055–4061.https://doi.org/10.1021/nn900748j

[47] J. H. Golbeck, C.Lubner, P. Khorzer, P. Silva, K. A. Vincent, Wiring an [FeFe]-hydrogenase with photosystem I for light-induced hydrogenproduction. Biochemistry.,49 (2010) 10264-10266. https://doi.org/10.1021/bi1016167

[48] C. A. Caputo, M. A. Gross, V. W. Lau, C.Cavazza, B. V. Lotsch, E. Reisner Photocatalytic Hydrogen Production using Polymeric Carbon Nitridewith a Hydrogenase and a Bioinspired Synthetic Ni Catalyst,Angew. Chem. Int. Ed. 53 (2014)11538-11542. https://doi.org/10.1002/anie.201406811

[49] P. D. Tran, A. Morozan, S. Archambault, J. Heidkamp, P. Chenevier, H. Dau, M. Fontecave, A. Martinent, B. Jousselme, V. Artero,A noble metal-free proton-exchange membrane fuel cell based on bio-inspired molecular catalysts,Chem. Sci., 6 (2015) 2050–2053. https://doi.org/10.1039/C4SC03774J

[50] S. Chandrasekaran, S. J. P. McInnes, T. J. Macdonald, T. Nann, N. H. Voelcker, Porous silicon nanoparticles as a nanophotocathode for photoelectrochemical water splitting,RSC Adv.,5 (2015) 85978-85982.https://doi.org/10.1039/C5RA12559F

[51] B. Kumar,M. Beyler, C. P. Kubiak, S. Ott, Photoelectrochemical Hydrogen Generation by an [FeFe] Hydrogenase Active Site Mimic at a p-Type Silicon/Molecular Electrocatalyst Junction, Chem. Eur. J., 18(2012) 1295 – 1298.https://doi.org/10.1002/chem.201102860

[52] T. Nann , S. K. Ibrahim, P. M. Woi, S. Xu,J. Ziegler, C. J. Pickett, Water Splitting by Visible Light: A Nanophotocathode for Hydrogen Production, Angew. Chem. Int. Ed. 49 (2010) 1574 -1577.https://doi.org/10.1002/anie.200906262

[53] M. Hambourger, M. Gervaldo, D. Svedruzic, P. W. King, D. Gust, M. Ghirardi, A. L. Moore, T. A. Moore,[FeFe]-Hydrogenase-Catalyzed H$_2$ Production in a Photoelectrochemical Biofuel Cell, J. Am. Chem. Soc., 130 (2008) 2015–2022.https://doi.org/10.1021/ja077691k

[54] G. Goldet, A. F. Wait, J. A. Cracknell, K. A. Vincent, M. Ludwig, O. Lenz, B. Friedrich, F. A. Armstrong, Hydrogen Production under Aerobic Conditions by Membrane-Bound Hydrogenases from *Ralstonia* Species. J. Am. Chem. Soc., 2008, 130, 11106-11113. https://doi.org/10.1021/ja8027668

[55] F. Y. Wen, X. L. Wang, L. Huang, G. J. Ma, J. H. Yang, C. Li, A Hybrid Photocatalytic System Comprising ZnS as Light Harvester and an [Fe$_2$S$_2$] Hydrogenase Mimic as Hydrogen Evolution Catalyst, Chem. Sus. Chem, 5 (2012) 849-853.https://doi.org/10.1002/cssc.201200190

[56] C. B. Li, Z. J. Li, S. Yu, G. X. Wang, F. Wang, Q. Y. Meng, B. Chen, K. Feng, C. H. Tung, L. Z. Wu, Interface-directed assembly of a simple precursor of [FeFe]–H_2ase mimics on CdSe QDs for photosynthetic hydrogen evolution in water, Energy Environ. Sci., 6 (2013) 2597-2602.https://doi.org/10.1039/C3EE40992A

[57] T. N. Huan, R. T. Jane, A. Benayad, L. Guetaz, P. D. Tran, V, Artero, Bio-inspired noble metal-free nanomaterials approaching platinum performances for H_2 evolution and uptake, Energy Environ. Sci., 6 (2016) 940-947.https://doi.org/10.1039/C5EE02739J

[58] F. Wang, W. Wang, X. Wang, H. Wang, C. Tung,L.Wu, A Highly Efficient Photocatalytic System for Hydrogen Production by a Robust Hydrogenase Mimic in an Aqueous Solution. Angew. Chem., Int. Ed., 50 (2011) 3193-3197.https://doi.org/10.1002/anie.201006352

[59] F. Peng, Q. Zhou, D. Zhang, C. LuY. Ni, J.Kou, J.Wang, Z. Xu, Bio-inspired design: Inner-motile multifunctional ZnO/CdSheterostructures magnetically actuated artificial cilia film for photocatalytic hydrogen evolution, Appl.Catal. B. Environ., 165 (2015) 419–427.https://doi.org/10.1016/j.apcatb.2014.09.050

[60] Y. Chen, J. He, J. Li, M. Mao, Z. Yan, W. Wang, J. Wang, Hydrilla derived $ZnIn_2S_4$ photocatalyst with hexagonal-cubic phase junctions: A bio-inspired approach for H_2 evolution, Cat. Commun., 87, (2016) 1-7.https://doi.org/10.1016/j.catcom.2016.08.031

[61] H. Zhou, X. Li,T. Fan, F. E. Osterloh, J. Ding, E. M. Sabio, D. Zhang, Q. Guo, Artificial Inorganic Leafs for Efficient Photochemical Hydrogen Production Inspired by Natural Photosynthesis,Adv. Mater. 2010, 22, 951–956. https://doi.org/10.1002/adma.200902039

[62] J. Lee, K. Yong, Combining the lotus leaf effect with artificial photosynthesis: regeneration of underwater superhydrophobicity of hierarchical ZnO/Si surfaces by solar water splitting, NPG Asia Mater 7 (2015)201. https://doi.org/10.1038/am.2015.74

[63] D. Nocera, The Artificial Leaf. Acc. of Chem. Res.,45 (2012) 767-776.https://doi.org/10.1021/ar2003013

[64] J. Sun, J. Zhang, M. Zhang, M. Antonietti, X. Fu, X. Wang, Bioinspired hollow semiconductor nanospheres as photosynthetic nanoparticles. Nat.Commun., 3 (2012) 1139.https://doi.org/10.1038/ncomms2152

Keyword Index

Artificial Photosynthesis.................... 211

Biogenic Synthesis 1
Bio-Inactivation................................. 39
Biomass 141
Biomimics... 211

Catalysis... 1
Charge Separation Mechanism............ 83

Disinfection 39
Dye-Sensitized Solar Cells (DSSCs). 175

Efficiency... 175
Electrolyte... 175
Energy Storage 141

Graphene... 141
Green Synthesis 175
Green Synthesis 39

Hydrogen Evolution Reaction 117
Hydrogenases.................................... 211

Light Irradiation................................. 39

Metal Nanoparticles.............................. 1
Metal Oxides...................................... 39
Microorganisms 39

Organic Pollutant Degradation............. 1
Oxygen Evolution Reaction 117
Oxygen Reduction Reaction............. 117

Photocatalytic Degradation 83
Photoelectrochemical Cell................ 211
Phytochemicals.................................... 1
Protein Nanotubes............................. 141

Quantum Dots................................... 175

Small Organic Molecule Oxidation
    Reaction...117

Supercapacitors 141
Synthesis...83

Toxic Chemicals..................................83

Water Splitting 211

About the Editors

Dr. Alagarsamy Pandikumar is currently working as Scientist in Functional Materials Division, CSIR-Central Electrochemical Research Institute, Karaikudi, India. He obtained his Ph.D. in Chemistry (2014) from the Madurai Kamaraj University, Madurai and then successfully completed his post-doctoral fellowship tenure (2014-2016) at the University of Malaya, Malaysia under High Impact Research Grant. His current research involves development of novel materials with graphene, graphitic carbon nitride, in combination to metals, metal oxides, polymers and carbon nanotubes for energy conversion and storage and dye-sensitized solar cells applications. His results outcomes were documented in 134 in peer-reviewed journals including 10 review articles and also have more than 3300 citations with the h−index of 39. On other side, he served as Guest Editor for a special issue in Materials Focus journal and edited 12 books for reputed publishers.

Dr. Perumal Rameshkumar is currently working as an Assistant Professor of Chemistry at Kalasalingam Academy of Research and Education, India. He obtained his M.Sc. (chemistry) (2009) from Madurai Kamaraj University. He joined as Junior Research Fellow (2010) at the same University and subsequently promoted as Senior Research Fellow (2012). His doctoral thesis focused on 'polymer encapsulated metal nanoparticles for sensor and energy conversion applications'. He worked as Post-Doctoral Research Fellow (2014) at University of Malaya, Malaysia in the field of 'graphene-inorganic nanocomposite materials for electrochemical sensor and energy conversion'. His current research interests include synthesis of functionalized nanomaterials, electrochemical sensors, energy-related electrocatalysis and photoelectrocatalysis. His research findings were documented in 34 peer reviewed journals including 01 review article. For his credit, he edited 02 books under Elsevier publications.

241

www.ingramcontent.com/pod-product-compliance
Lightning Source LLC
Chambersburg PA
CBHW071159210326
41597CB00016B/1603